THE ART OF BEING A **PARASITE**

THE ART OF BEING A PARASITE

CLAUDE COMBES

Translated by

Daniel

Simberloff

The

University of

Chicago

Press

◊

Chicago and

London

Claude Combes
is professor of animal
biology and director of
the Centre de Biologie
et Écologie Tropicale et
Méditerranéenne at
the Université de
Perpignan, France. He
is the author of
*Parasitism: The Ecology
and Evolution of
Intimate Interactions.*

Daniel Simberloff
is the Nancy Gore
Hunger Professor of
Environmental Studies
at the University of
Tennessee. He is coedi-
tor of *Strangers in
Paradise: Impact and
Management of
Nonindigenous Species
in Florida* and transla-
tor of *Killer Algae* by
Alex Meinesz.

The University of Chicago Press, Chicago 60637
The University of Chicago Press, Ltd., London
© 2005 by The University of Chicago
All rights reserved. Published 2005
Printed in the United States of America

14 13 12 11 10 09 08 07 06 05   1 2 3 4 5

ISBN: 0-226-11429-5 (cloth)
ISBN: 0-226-11438-4 (paper)

Also published in French as *Les associations du vivant:
L'art d'être parasite,* © Flammarion, Paris, 2001.

Library of Congress Cataloging-in-Publication Data

Combes, Claude.
[Associations du vivant. English]
The art of being a parasite / Claude Combes;
translated by Daniel Simberloff.
p. cm.
Includes bibliographical references and index.
ISBN 0-226-11429-5 (cloth: alk. paper) —
ISBN 0-226-11438-4 (pbk. : alk. paper)
1. Parasites. 2. Parasitism. I. Title.
QL757.C614513 2005
577.8'57—dc22
2005000674

CONTENTS

What Is a Symbiosis?

The Tree of Life and Genetic Information

We are living and surrounded by life.

Listen to Prévert: "Then all the animals, the trees and plants, begin to sing, to sing, to sing at the tops of their voices, the true living song, the song of summer, and everyone drinks, everyone raises a glass."[1]

Everyone . . .

Prévert was able to say "everyone." So many different living beings, those that are visible and those that are unseen. And some of these unite to form symbioses, or living associations. Because one of the fundamental characteristics of life is that it is present on our planet in discrete forms. But the discreteness can sometimes be erased. We can imagine that life arose on our planet just once and that it remained in the form of a single species, evolving or not throughout the geological ages. Possibly the first of these propositions (that life arose just once) is correct. The second is obviously false.

Far from remaining monotonous, life has exploded into a multitude of distinct species. These species are separated from one another by the criterion of reproductive isolation: the horse and the cow do not hybridize; thus, they belong to separate species. The horse and the donkey mate, but their hybrids are sterile; therefore they also belong to separate species. We believe that several million species (of which only 1.5 million are known) constitute the current biosphere and that possibly 2 billion species have existed at one time or another since the origin of life about 4 billion years ago.

Parasites form a large proportion of life on the earth.

◊

PETER W. PRICE (1980)

2   In the context of biological evolution, this splitting of life into a multitude of species can be represented as an enormous tree whose shape we try to reconstruct by various means. Each node or branch point of the tree corresponds to an act of speciation, the separation of a parent species into two daughter species.[2] This tree can be interpreted in terms of information.

Each living being, from the most primitive in terms of complexity to the most highly evolved (which is usually considered to be humans), is constructed on the basis of information encoded by the nucleic acids of the genotype.[3] Because they are structured as a succession of nucleotides, nucleic acids can encode blueprints of a nearly infinite number of proteins.[4] Starting with these proteins, cascades of interactions lead to the synthesis of many other molecules and finally to the construction of an entire organism, whose physical manifestation is known as the phenotype. For example, the construction of a human being requires some three billion nucleotides (although only a fraction of them actually engage in making proteins).

When speciation occurs—that is, when an ancestral species gives rise to two daughter species that cease to interbreed—the information associated with each daughter species is definitively isolated from that of the other species, just as branches of a tree do not rejoin once they have separated.

Associations, however, can form between species that have followed separate evolutionary trajectories (often for a very long time). Let me return to the metaphor of evolution as an immense tree: it is almost as if two branches come together and from then on are associated. Such associations are termed *symbioses* (fig. 1). In the great majority of cases, one of the two species uses the other not only as habitat but also as food. The species inhabiting the other is the parasite, and the species it inhabits is the host. As we shall see, symbioses are generally strongly asymmetric.

Of course there exist many more nuanced sorts of symbioses; the most important variant is when the exploitation is not always in the sense I have just described. In some cases (the number is increasing as study of the phenomenon becomes more intensive) it is not the inhabitant species that is exploiting the habitat species, but the inverse. To speak in deliberately provocative terms, I can say that it is not the parasite that exploits the host, but the host that exploits the parasite. Because this inversion cannot occur unless there is at least some reciprocity in the exchange of benefits between species, such symbioses are called *mutualisms* rather than *parasitisms*. I return to this apparent paradox at the end of this chapter.

In a symbiosis, whether a parasitism or a mutualism, the genetic information of each species can interact with that of the other in two ways. First, each species may contribute to some degree in the production of the phe-

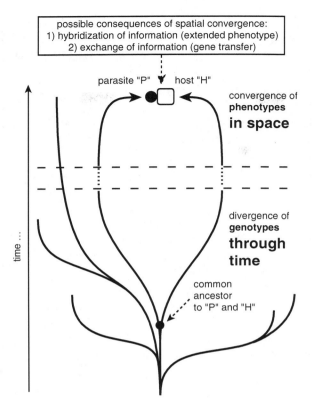

possible consequences of spatial convergence:
1) hybridization of information (extended phenotype)
2) exchange of information (gene transfer)

parasite "P"   host "H"

convergence of **phenotypes** **in space**

divergence of **genotypes** **through time**

common ancestor to "P" and "H"

time ...

Fig. I. Hypothetical schema showing how two repositories of genetic information, separated for millions of years, can find themselves associated in a symbiosis.

notype of the other species, thus extending its genetic expression into the phenotype of its partner. This is a hybridization of information. Second, these species can extend the interaction still further by exchanging DNA sequences. This is an exchange of information.

We will see throughout this work that symbioses, whether parasitic or mutualistic, have played a key role at several points in evolutionary history.

The Hybridization of Information

By *hybridization of information* I mean what Richard Dawkins (1982) has called, in a striking metaphor, the *extended phenotype*. What do I mean? In response to this question, let us consider, with Dawkins, galls that certain parasitic insects induce in their host plants.

When, for example, a wasp in the family Cynipidae lays an egg on the leaf of an oak or a rose bush, cells of the leaf multiply to form a mass around the developing parasite. This excrescence is the gall. Its apparent "function" is

4   twofold: on the one hand it protects the parasite from various enemies by enclosing it in a thick wall; on the other hand it provides nourishment, because the insect larva consumes the gall from within it. Moreover, the insect actually controls the growth of the gall in that the structure of this excrescence arises gradually, apace with the needs of the larva.

It is clear that, in the absence of the insect, there would be no intense cellular multiplication of the leaf at the spot where the gall has formed; it is the presence of the parasite that modifies the plant phenotype. The genome of the wasp parasite is expressed in the phenotype of its host plant. However, the hybridization of information is two-way; I will show that the genome of the host can be expressed in the parasite.

We should conclude from this interaction that a genome can influence cells, tissues, and organs that it has not constructed itself, that were constructed by another genome. Most often, as with galls, this extension of the parasite phenotype into the host phenotype is to the advantage of the parasite. It results from natural selection in the sense that individual parasites carrying mutations (that have arisen by chance) that give them an adaptive advantage will reproduce more than other parasites, thus increasing the frequencies of these mutated genes in the parasite population.

In the course of this work, we will see other examples of hybridization of information. We will learn in particular that the parasite's information can provoke changes in the host that favor either the transmission of the parasite to its host (for example, by altering the behavior of intermediate hosts in complex life cycles) or the survival of the parasite that has reached its host (for example, by inducing immunosuppression).

The host can also express its genes in the parasite's phenotype. For example, all the defense mechanisms that limit the multiplication of the parasite are aspects of the host's extended phenotype.

The Exchange of Information

The fact that an organism lives permanently on or in another organism does not imply that their genotypes fuse. To the best of our knowledge, it seems that such fusion does not occur. Rather, partial exchanges of genomes can take place in the form of longer or shorter fragments of nucleic acid molecules that pass from the genome of one partner to the genome of the other.

Our conception of the genome, with respect to its fluidity, greatly evolved in the second half of the twentieth century. When the hereditary elements were conceived as discrete units called genes, an agreed-upon image arose: the genes were aligned in orderly fashion, one behind the other on the chromosomes. This image still seems largely correct, except for the word *orderly*. Far from behaving in orderly fashion, DNA sequences are capable of many tricks. They can double themselves, lose pieces of themselves, and especially jump around, either from one part of a genome to another part of the same genome or to another genome altogether.

For instance, consider the functioning of retroviruses such as HIV,[5] which causes AIDS.

Schematically, when a human lymphocyte (white blood cell) is in contact with a particle (the virion) of HIV, an adhesion is quickly produced between a molecule on the surface of the virion and several molecules on the surface of the lymphocyte (receptors and coreceptors). This adhesion is followed by the entrance of the viral genome, which at this point is composed of RNA. In the lymphocyte, the virus subverts the metabolic machinery of the lymphocyte so that it constructs an enzyme, reverse transcriptase, that transcribes the viral genetic information into a DNA sequence. This DNA molecule undergoes various transformations and then penetrates the nucleus of the lymphocyte. There the viral DNA inserts itself into the lymphocyte DNA, just as a dancer squeezes between two others in a farandole. It's easy to guess the consequence: the virus in this integrated form uses the replicative machinery of the host genome to produce many replicas of its genetic information and, at the same time, of the proteins it needs to construct virions. These virions leave this cell and infect other cells.

It is the integration of the viral DNA into the host DNA that is the crucial event. It shows us in effect that it is possible for a nucleic acid sequence to insert itself into another nucleic acid sequence. Now, in a symbiosis such as a host–parasite system, the promiscuity of the two genomes, strangers but spatially (and persistently) very close, is conducive to this sort of exchange.

We know today that all genomes are invaded by parasitic sequences, known as mobile elements, that in general have the same sorts of properties as the retrovirus HIV, even if the biochemical processes that allow them to multiply and to move are not identical. We also know that these mobile elements can capture pieces of DNA belonging to the host and carry them. The stage is then set for unrelated genomes to exchange information by virtue of the mobile elements or other means. Does this mean this phenomenon is a common occurrence? The answer is no.

Genetic exchanges between a parasite and its host probably happen rarely in most host–parasite systems, but the important fact is that they do happen. In fact, the older the host–parasite association, the greater the likelihood that

6   such exchanges have occurred. And some symbioses are so old (tens or even hundreds of millions of years) that exchanges have occurred in many of them. For example, in the association of the mitochondrion—which is nothing but an ancient bacterium transformed into a mutualist—and the eukaryote cell in which it is found, the reality of genetic exchanges has been amply demonstrated by molecular means (see Selosse, Albert, and Godelle 2001). It is true that the association between mitochondria and eukaryote cells is at least a billion years old!

Who Is Associated with Whom?

In order for either a hybridization of information or an exchange of information to occur, there must be at least two partners whose genomes initially differ. It is worth noting in passing that the phrase *at least* is not simply stylistic. Even if most research is focused on pairs of species in order to facilitate understanding, the nature of some symbioses implies many more than two partners. A human individual, for example, is associated with (1) DNA sequences that are apparently part of his or her genome but that have, in fact, a foreign origin in the more or less distant past; (2) active or dormant viruses such as that of chicken pox that remain in the body throughout one's life (but not in the DNA!) and can activate to cause diseases such as shingles or herpes; (3) mitochondrial bacteria, already mentioned, which have a fundamental role in the metabolism of every cell; (4) many bacterial species in the intestinal "flora," some of which are indispensable; and (5) various parasites, ranging from the intestinal pinworm that is hardly pathogenic to the malarial plasmodium that can infect up to a third of an individual's red blood cells.[6] We see that this list, reminiscent of Prévert's song, includes organisms that are extremely diverse by virtue of their taxonomic positions and their effects on the host.

Which Are the Most Common Partners?

For the hosts, the answer is easy: all living organisms (even the smallest, like viruses) can be hosts and therefore can house parasites or mutualists. The only limit is size: the smaller an organism, the more limited the list of parasites it can support. For example, a virus can harbor only nucleic acid molecules smaller than itself. As for the parasites, the answer is more complicated. Parasites can be sequences of nucleic acids, viruses, bacteria, unicellular or multicellular organisms, plants, fungi, or animals.

Nowadays we do not differentiate between viruses and bacteria on the one hand and parasites in the more traditional sense on the other. Among the entire gamut of parasitic organisms, the only distinction used today is that between microparasites (viruses, bacteria, fungi, and protists) and macroparasites (helminths, arthropods, and other metazoans). This distinction is based on several biological characteristics.

In general, *microparasites* are small, multiply profusely on their hosts, induce a lasting immunity, have unstable populations, and as a result cause epidemic diseases. In general, *macroparasites* are larger, do not multiply on the hosts (with some exceptions), do not induce lasting immunity (here also, there are exceptions), have more stable populations, and therefore cause endemic diseases.

Among the different groups of living organisms, the "passage to a parasitic existence" has greatly differing frequencies. To cite just three examples from among the metazoans: most platyhelminths (flatworms) are parasitic, whereas among nematodes (roundworms) there are about equal numbers of parasitic and free-living species, and there are no parasites at all among echinoderms (sea urchins and starfish).

The most important observation is that there is no rule whatsoever as to the relationship between partners in a symbiosis. It is as likely that the genetic distance is enormous (as between a virus and its vertebrate host) as that it is very small (parasitologists know many cases of parasitism between close relatives). The most frequently cited example of the latter phenomenon is that of many red algae that parasitize other, closely related red algae. We will even find (chap. 3) an example of parasitism between two wasps belonging to the same genus (*Polistes*). If we extend the concept of parasitism to all cases of a lasting exploitation of one organism by another, we would not hesitate to classify as intraspecific parasitism well-known cases among several bird species (swallows, starlings, moorhens, and others; we return to such species in chap. 6) in which females deliberately lay several eggs in the nest of a neighboring pair. And one would surely be licensed to study "social parasitism" among humans, the subject of several generations of social scientists.

Arms Races

The Leitmotif of Life

The study of symbioses nowadays is always undertaken from a modern evolutionary perspective, that of Darwinism as it has been remodeled throughout the twentieth century and finally illuminated by discoveries in molecular biology. It is therefore important to recall that the significant thing in evolution is the reproductive success of individuals (or of populations), often termed *fitness*. If, for example, a pair of tits that lay eggs in a "clean" nest rear on average ten nestlings to fledging, whereas a pair that lay their eggs in a flea-ridden nest can raise only five offspring, one can say that parasitism by the fleas has lowered the reproductive success, or fitness, by half. (The real impact would be still greater if the offspring of the infested nest are of lower "quality" and therefore tend to survive for shorter periods.) In these terms, an advantage or benefit is thus augmented reproductive success, while a disadvantage or a cost is diminished reproductive success. Natural selection is the common thread that knits together all of evolution.[1]

To transmit their genes to the next generation appears as the leitmotif of all species that have existed on earth since the origin of life. When Charles Darwin enunciated the key principles of this process, he caused a scandal for several reasons. The main one was that without doubt he implicitly based all evolution (and right at the outset the origin of humans was at issue) on processes deprived of any moral or ethical content. People knew that wolves ate sheep and lions ate gazelles, but what could pass for a subsidiary clause of the punishment of Adam and Eve suddenly became, because of Darwin, a central mechanism of life. What am I saying? The central mechanism of *all life!*

Nearly a century and a half later, despite many discoveries, the principle of natural selection has never been cast into doubt. As for symbioses, they offer an excellent opportunity to discover and understand the workings of natural selection. If one imagines a caricature of a population of hosts confronted with a population of parasites, the hosts that are best at transmitting their genes to the next generation are those that defend themselves best against the parasites. Similarly, the parasites that transmit their genes best to the next generation are those that exploit their hosts best. We will see later that reality is not so simple but that this is a good description of the basic operation of natural selection. It implies reciprocal selective pressures. In a sense, the hosts select for the best parasites and the parasites select for the best hosts.

Encounter and Compatibility

We often compare predator–prey systems with those of parasites and hosts. Similarities do exist: in both systems, the interaction occurs only if there is an encounter. Examples are the encounter between a mouse and a cat in a predator–prey system and the encounter between an infective stage of a parasite and a mouse in a host–parasite system. But there follows a major difference between the two types of systems, as shown in figure 1.1.

In the predator–prey system, there is no post- encounter interaction. The mouse is eaten as soon as it is captured. What remains in the stomach and then the intestine of the cat during the brief period of digestion is no longer the mouse (and still less, the information that the DNA of the mouse carries) but common molecules from the cells that had made up the mouse.

By contrast, we can say that in the host–parasite system, the real action of the interaction actually begins after the encounter: either the mouse is able to destroy the infective stage of the parasite or the parasite is able to survive in the mouse. If the parasite survives, it manages to install itself in the right microhabitat and reproduces. This state can last for weeks or months, even for years. This second phase in the host–parasite relationship, after the encounter, is that of compatibility. We say there is a lasting, intimate interaction between the two partners (Combes 2001).

It is this lasting interaction that allows the hybridization and exchange of information that I discussed earlier.

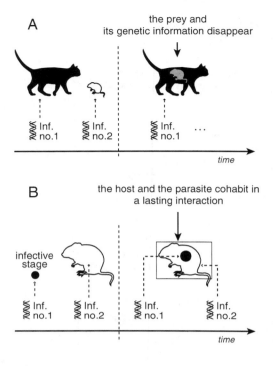

Fig. 1.1. The fate of genetic information in a predator–prey system (A) and a host–parasite system (B). Only in the latter are the two sets of genetic information conserved side by side. This is a persistent, intimate interaction. Inf., information.

It is easy to see that from an evolutionary point of view it is in the best interest of the parasite to possess adaptations that allow it (1) to *encounter* its host and (2) to *survive* in the host if the encounter has occurred. Conversely, it is in the best interest of the host to have adaptations that allow it (1) to *avoid encountering* the parasite and (2) to *get rid* of the parasite if, despite any efforts to avoid an encounter, one has taken place. From these considerations arises an "arms race," an expression that evokes the reciprocal selective pressures that the parasite species and the host species exert on one another's evolution over long periods of time, even millions of years.

Let us imagine a host–parasite system at any time during its evolution and suppose that the parasite population is genetically diverse, so that certain individuals in particular infective stages behave in a genetically determined way that gives them a higher probability of encountering a host individual. Clearly, parasites that possess this behavioral trait will be positively selected, and their genes will tend to increase in frequency from generation to generation in the parasite population. For example, the behavior in question might be a positive response to a stimulus (odor, vibration, etc.) emanating from the host.

Now let us consider the host and imagine that certain individuals behave in a genetically determined way that reduces the probability that they will encounter an individual of the infective stage of the parasite. If the parasite is even slightly pathogenic, these host individuals that have avoided the parasite will be healthier than infected individuals. On average, they will be more successful reproductively, and their genes will tend to increase in frequency in the host population. The behavior that helps them avoid the parasite might be, for instance, fleeing to a habitat that the infective parasite stage cannot reach.

We see that these two selective mechanisms, one operating in the parasite population, the other in the host population, have all the features of an arms race. The more successful the parasites are in finding hosts, the more intense will be selection on the hosts, and the more beneficial to the host will be any adaptations that help it to avoid the parasite. The two species have thus engaged in an endless process. To the extent that the host becomes more effective at avoiding the parasite, the parasite survives only if it has sufficient genetic diversity that natural selection can produce better means of encountering the host. In turn, of course, selection in the host replies by producing new ways to avoid the parasite. In other words, selection for "encounter genes" in the parasite genome generates selection for "avoidance genes" in the genome of the host, and vice versa (fig. 1.2, *left*).

If we were dealing with a predator–prey system, things would stop here. The process I have just described can easily be applied to the relationship between cats and mice or between lions and gazelles. The cats and the lions

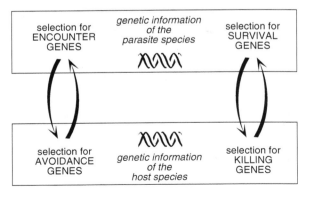

Fig. 1.2. The two arms races in a host–parasite system: genes to facilitate encounters vs. genes for avoidance (*left*), and genes for survival vs. genes for efficient killing (*right*).

have genes that help them encounter the mice and the gazelles, respectively. The mice and gazelles, in turn, have genes that help them avoid cats and lions. But in a host–parasite system, the story does not stop with the encounter.

If an encounter has occurred, the host still has a chance to get rid of the parasite. The host possesses the astonishing property of being able to transform itself from the game to the hunter! All species (not only vertebrates but even the most primitive invertebrates) are able to recognize as foreign (that is, not-self) all molecules or ensembles of molecules that did not arise within them and to deploy against these intruders a battery of weapons. In the most highly evolved species, these weapons are extraordinarily elaborate, involving several sorts of cells (those that produce antibodies, those that are cytotoxic, etc.) and circulating molecules that are equally diverse (antibodies, cytokines, etc.).

The immune system mechanisms exert formidable selective pressure against pathogenic agents. The only parasite individuals that transmit their genes to the next generation are those whose traits allow them to survive in spite of the hostile milieu created by the host. Obviously, the selection of parasites able to survive maintains pressure on the host, so that natural selection allows host individuals to survive who have new weapons. All the ingredients are in place for a second arms race. The better the host can struggle against the pathogens, the more the pathogens are forced to adapt to the host's armaments. Conversely, the better the parasite is able to cope with the host's armaments, the more strongly natural selection forces the host to acquire new ones. The selection in the genome of the parasite for ability to survive in the host entrains selection in the host genome for ability to kill the parasite, and vice versa (fig. 1.2, *right*).

How the Arms Races Work

The two arms races I have just described can be represented by two filters—one for encounter, the other for compatibility. These two filters can be drawn as diaphragms (fig. 1.3). Natural selection operates in the parasite genome to open the two filters, and it operates in the host genome to close the two filters. What do these filters represent, from a genetic point of view?

These are *hybridized phenotypes* (Combes 2001), because their status at any moment depends not just on the genes of the parasite or on the genes of the host but on the genes of both individuals. These hybridized phenotypes obey the following rules.

1. The degree to which the filters are open determines not only the possibility or impossibility that parasitism will occur but also the abundance of

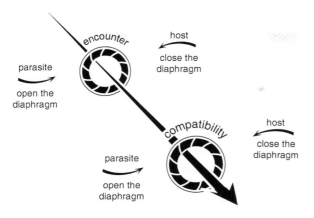

Fig. 1.3. Encounter and compatibility in host—parasite systems, represented by two diaphragms. Selection in the parasite genome tends to open the passage, whereas selection in the host genome tends to close it. The filters are hybridized phenotypes, undergoing opposing selection pressures in the parasite and host.

the parasites if it does. This means there is very little chance that the degree of opening will remain constant. On the contrary, encounter and compatibility are both strongly influenced not only by eventual new mutations but also by *environmental factors*. For instance, a rainy year can favor completion of the life cycle of the large liver fluke because pastures are more humid (thus transmission is more frequent). Or an increase in the density of sheep can entrain a nutritional deficit and a consequent lowering of immune defenses (thus compatibility increases).

2. One closed filter alone suffices to keep the host from being parasitized. This fact confirms that hosts possess two sequential lines of defense and that, depending on the specifics of each particular case, selection can act more strongly on behavior that keeps the host from encountering the parasite or on immunity that protects the host after it has been parasitized. These alternatives Michael Hochberg (1997) has characterized by the phrase "hide or fight?"

3. The two arms races—of encounter and compatibility—are not independent in the sense that the adaptive responses are interchangeable. For example, to an increase in the frequency of encounters generated by natural selection acting on the parasite genome, the host can respond by increasing the efficiency of its immune system. The initial degree of parasitism would then be maintained.

As I have just described, host–parasite systems differ from predator–prey systems in one crucial detail. When a cat chases a mouse, both individuals run, but their goals are not the same. The cat runs in order to

14 get a meal (after all, the cat will still be able to survive if it does not catch this particular mouse). The mouse, however, is running for its life (Dawkins and Krebs 1979). When an individual of an infective stage of a parasite seeks to infect a host individual, they both run (at least symbolically), but in this case it is the parasite that is running for its life, because it will die if it does not quickly find a suitable host. The host runs for its life only in particular cases; in general, it runs in order to be healthier.

This detail allows us to understand why there always are mice and parasites. The mice possess enough adaptations (to detect predators, to run quickly, to hide in inaccessible places, etc.) that a sufficient fraction of them escape cats. For their part, parasites possess enough adaptations both to encounter hosts (invisible infective stages, insertion in the food chain, use of vectors, etc.) and to survive even highly developed defense mechanisms (molecular mimicry, antigenic variation, immunosuppressive ability, etc.). Of course many infective-stage individuals die because they do not find a host, and many of those that do find hosts are killed after they infect them. Here also, however, enough parasites escape the slaughter, and the symbiosis persists from generation to generation.

Arms races also exist in mutualistic systems, even in obligatory mutualisms in which each of the two partners cannot survive without the other. Each partner (I return to this debate in chap. 5) remains fundamentally self-interested, and conflicts can linger for a very long time. Thus there are still conflicts between the mitochondrial genomes and nuclear genomes of eukaryote cells, even though this association is one of the oldest known. The only difference is that conflicts between mutualists are less severe (and thus less visible) then those between true parasites and hosts.

The health sciences today know ways of influencing just how open these filters are. Closing the encounter filter is achieved by education and hygiene measures that lead to the avoidance of infective stages and contagion. As for closing the compatibility filter, this is achieved in two ways: (1) by therapy (for example, by antibiotics, either synthetic or natural) that suddenly renders the internal milieu of the host (that is, the habitat of the parasite) hostile for pathogens; or (2) by vaccination, which renders this milieu hostile even before the parasite invades, by virtue of priming the host immune system to respond immediately upon invasion.

The struggle against parasites can be compared with the manner in which cities protected themselves in the Middle Ages (fig. 1.4). In both instances there are two lines of defense. Either one prevents contact with the enemy (by closing the encounter filter), or if the enemy penetrates, one engages in hand-to-hand combat (by closing the compatibility filter).

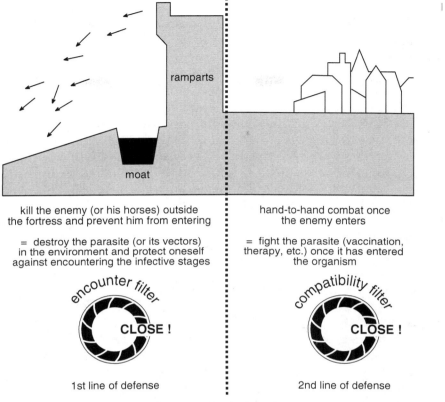

Fig. 1.4. The defenses of the human individual against parasites, compared with the defenses of a fortress (vaccination is equivalent to training soldiers).

Certain parasites themselves close the compatibility filter to subsequent invaders. For example, we contract measles only once in a lifetime. In an unvaccinated individual, the compatibility filter is wide open when the virus is first encountered, but it closes as infection proceeds and remains closed for the rest of the individual's life. In some systems, the filter can reopen after a certain period. This is the case for chicken pox: the virus that became dormant after a child's infection can become active again when the child grows up, breaking out as shingles. These phenomena are observed primarily with viruses. By contrast, they are rarely effective in multicellular parasites.

(Nearly) Invulnerable Parasites

Hosts appear to have very solid defenses with which to oppose intruders of all sorts that seek to exploit them.

First, the amount of DNA (DNA coding for proteins) in organisms such as vertebrates generally greatly exceeds that typical of parasites. The human genome contains probably ten times as many genes as does the genome of the agent of schistosomiasis and a hundred times as many as does the genome of the agent of malaria. The basic genetic resources are thus greater in hosts than in parasites.

The genetic resources are amplified still further, at least in higher vertebrates, by what are called rearrangements of the genome. Although certain molecules used by higher vertebrates to struggle against pathogens are produced by a single gene, the majority, by contrast, result from the fragmentation of genes, followed by splicing associated with chance rearrangements of the fragments that create new sequences. The number of combinations that can be obtained this way is enormous—on the order of several tens of millions of combinations with several tens of underlying genes. This is true notably of the antibodies that humans direct against all sorts of "non-self" entities that might invade us. The regions of the genome involved in this defense are also the sites of other mechanisms that increase the diversity of responses still more. Among these mechanisms are many mutations, including those that entail adding a single nucleotide; such a mutation is called a frameshift because it shifts the reading frame and creates new codons throughout the remainder of the DNA molecule.

Overall, however, the parasites possess several decisive assets in the arms races. A process of selection implies that there exists a continuous renewal of genetic variation by mutation, and it is here that the match is highly unequal between parasites and hosts. Mutations that will be transmitted to offspring can occur only during meiosis, when gametes form. Three key factors give parasites their advantage: generation length, mutation rate, and fecundity.

The shorter the generation time, the more numerous are the occasions (meioses) for mutations to be produced in the cells of the "germ line" (gamete-producing cells). The smaller the parasites (and they are always smaller than their hosts), the shorter their generation times. Ridley (1994) observed, for example, that, if (on average) bacteria divide every thirty minutes and humans reproduce every thirty years, a human lifetime contains nearly 1.5 million generations of bacteria, whereas there have been only 200,000 generations of primates since the hominid line split from the chimpanzee line seven or eight million years ago.[2] We can imagine the degree to which the number of mutations (and therefore the rate of adaptation) of bacteria exceeds that of humans.

Moreover, the mutation rates themselves are very different, much higher in viruses and bacteria than in eukaryotes. In bacteria, there even exist mutator genes—genes that cause a great increase in mutation rate in certain circumstances. Among all the mutations that arise, many are lethal (entrain the death of the cell) or have no effect, but it takes only a few favorable mutations to cause the rapid selection of strains with some new property. This increase in the mutation rate occurs when some change in the environment stresses the bacterial population. It operates by a change in the repair systems for DNA. When a DNA molecule replicates, errors arise that modify the genetic information if they are not corrected. There are particular enzymatic proteins that detect the errors and usually repair them. In bacteria with mutator genes, these proteins botch their job under stressful conditions and many novelties appear, so the probability increases that by chance some traits will arise that are advantageous for surviving the stress (Taddei, Matic, and Radman 2000).

Finally, fecundity (I am using the word in a very broad sense here) differs greatly between the majority of parasites and their hosts. The common tapeworm of humans (*Taenia saginata*) produces up to 10 billion eggs during its lifetime. What would be the egg production of tapeworms of whales, *Polygonoporus* and *Tetragonoporus*—the longest extant animals known (40 meters)—which comprise tens of thousands of segments, each of which contains several male and female reproductive apparatuses? Of course a huge production of offspring does not imply that the few individuals that survive do so in general because of natural selection. Chance is a powerful contributor to their demise. But it is nevertheless true that, in all instances in which selective pressures can operate, they find abundant and genetically diverse material to work on among parasites.

I should add that in certain instances there has been selection in parasites for adaptive behavior that goes well beyond simple exploitation of a normally stable habitat. What can a parasite do when faced with a habitat that turns out to be unstable? Jovani (2003) discusses ectoparasites (lice and mites) that live amid bird feathers and are therefore periodically confronted with a molt. A molted feather constitutes a mortality factor for ectoparasites ensconced in old feathers. In fact, parasites can avoid feathers that will soon be molted and attach themselves to newly developing feathers. Such parasite behavior resembles that of free-living organisms that are able to move when environmental conditions become unfavorable.

Discussion: Who Exploits Whom?

In this chapter, I have drawn a distinction between parasites and mutualists. In fact, this distinction is quite arbitrary. To convince ourselves of this fact, we need only draw a parallel with human society:

- From a master–slave relationship, the master draws only advantages and the slave only disadvantages. In other words, the master would live more poorly without the slave and the slave would live better without the master.
- In a perfectly equitable relationship between a baker and a pastry chef, both profit from increases in prestige and clientele. Separately, they would attract fewer customers and lower sales.

Between the two types of association, a range of intermediates—that is, more or less equitable associations—exists. One can imagine that the fantastic diversity of life means that, as soon as associations arise between individuals of different species (and not just individuals of the same species), the distinction between parasitism and mutualism can become a headache. D. C. Smith (1992) even suggests that we no longer speak of parasites and mutualists but of hosts, symbionts, and exploitation.

It is necessary first of all to emphasize that Smith's reasoning is based on the notion of habitat. He distinguishes, just as I did earlier (see fig. 1 in the introduction), between an inhabitant and an inhabited entity. The inhabited entity is called the host, while the inhabitant is called the symbiont. Then he asks a remarkably simple question: Who is exploiting whom? His answer is succinct: Some hosts are exploited by their symbionts, and some hosts exploit their symbionts. Does this simplifying proposition apply to all symbioses?

The notion of hosts exploited by their symbionts encompasses the classical conception of parasitism. When a cow (the host) shelters in its bile ducts hundreds of liver flukes (symbionts), it is clear that the host is exploited by the symbionts without receiving anything in return.[3] The flukes behave like squatters who pay no rent.

The notion of hosts exploiting their symbionts is less evident, as is clear from a concrete example. In the oceanic depths exist unusual benthic communities. They are located at depths that no photon of light reaches, so there is no photosynthesis of organic matter mediated by chlorophyll. However, varied multicellular organisms live there, including worms (*Riftia*), bivalve molluscs, small crabs, and even some fishes.

One might ask, because sunlight never reaches these depths, where these organisms get their energy. The answer: from a chemical reaction, oxidation of sulfur-containing materials, which are common in hot springs that flow in some parts of the oceans. The problem is that no animal listed has genes that enable it to make a living by oxidizing sulfur-containing molecules. On the other hand, such genes exist among bacteria known as chemosynthetic.

By becoming hosts of these microorganisms, more highly evolved species can survive in the ocean depths. For example, whereas coastal bivalves (mussels, oysters, etc.) draw their energy from consumption of phytoplankton or zooplankton captured by their filtration apparatuses, bivalve molluscs living in deep hydrothermal vents, such as species of *Calyptogena* and *Bathymodiolus*, get their energy from symbiotic bacteria.[4]

What exact role does each of the two symbionts play in the bivalve–autotrophic bacteria systems? The host is the bivalve, because it shelters the vastly smaller bacteria in its tissues. The symbiont, in Smith's sense, is therefore the bacteria. Is the host exploiting the symbiont here or vice versa? It is easy to answer that it is the host that exploits the bacteria, because the bivalve's cells digest some of the chemosynthetic bacteria and the bivalve draws all its energy from this "predation." It should be noted that the sulfide-oxidizing bacteria can live alone, whereas these molluscs (*Calyptogena* and *Bathymodiolus*) are completely unable to do so.[5]

Of course it is this sort of association that has long led specialists to talk about mutual benefits rather than exploitation. For instance, they invoke the facts that the bacteria are protected in the mollusc tissues and that the molluscs provide the bacteria with sulfides to show that the host is helping the symbionts.[6] Certainly they are! But this sort of help exists in all exploitation systems simply because the exploiter has to manage the exploited individual: the deepwater bivalve shelters and feeds its bacteria in the same way that a master shelters and feeds his slave, as the individual who raises rabbits shelters and feeds his rabbits. But it is obviously the bacteria who do the work by transforming the raw nutrients into organic matter that the bivalve can assimilate.

A comparison with raising rabbits is even more justified because the two systems function, in terms of costs and benefits, in a completely parallel manner: the farmer provides the rabbits with shelter, the rabbits eat the grass the farmer provides for them, and the farmer eats the rabbits while conserving enough of them to keep the entire system from collapsing. The *Calyptogena* and *Bathymodiolus* shelter the bacteria, while the bacteria oxidize the sulfides that the bivalves procure for them. The bivalves eat the bacteria, but they leave enough of them in their tissues to assure their own futures, so to speak.

The examples of associations that I have just cited (the cow and its liver flukes, the deepwater bivalves and their bacteria) well illustrate that Smith is basically correct. Although the word *parasitism* does not cause problems, the word *mutualism* is deceptive. In mutualistic symbioses there is only an apparent mutualism, if this latter word is used in its strict sense.[7]

Does this fact mean I should abandon in this book the distinction regularly drawn between parasitism and mutualism? Certainly not! I use the word *host* in every case in which one of the partners of the symbiosis serves as a habitat for the other, whether the partner that is the habitat is the exploiter or the exploited. I use the word *parasite* when it turns out that the partner inhabiting the host damages it, and the word *mutualist* when the benefits are, or at least appear to be, quite equally shared between partners. The word *symbiont* can designate a parasite as well as a mutualist.

I should note, as Darwin discreetly implied and as Richard Dawkins (1982) loudly proclaimed, that genes are egoists. How could the phenotypes arising from these genes (the cow, the liver fluke, the bivalve, the bacterium, but also you and I) not be?

How Does One
Become a Parasite?

The Transition to Parasitism in the Course of Evolution

To confront a living environment as parasites do is not only to colonize a new type of environment but to begin an entirely new lifetime adventure. This living environment is the first environment that will defend itself. As I have shown in the preceding chapter, there exist highly varied tendencies toward a parasitic existence in different groups of species. The study of parasitism yields many surprises.

The phylum Coelenterata contains one of these surprises, for it includes some 10,000 free-living species (including the green hydras of rivers, the corals of warm seas, and jellyfish) and a single parasitic one.[1] This lone passage to parasitism is itself an enigma, but the strange nature of this passage is still more enigmatic because the species parasitizes caviar. This parasite, *Polypodium hydriforme*, spends part of its life living freely among plankton and another part in the oocytes of sturgeon (which, when they are healthy, produce a bounty for gourmet tables). A biologist at the Zoological Institute of St. Petersburg, E. V. Raikova, has spent the better part of her career elucidating the life cycle of this parasite. I summarize this research here (see esp. Raikova 1994).

P. hydriforme was discovered in 1871 in oocytes of sturgeon in the Volga River and was then found in several other sturgeon species, including the beluga (*Huso huso*). Although it has been found most

We don't doubt that *Tyrannosaurus* used its diminutive front legs for something. But . . .

◊

STEVEN J. GOULD *AND* RICHARD LEWONTIN (1979)

frequently in Russia, it is also known in Romania and Iran, and even in the state of Missouri in the United States (where it parasitizes species of the genus *Polyodon*, related to sturgeon but separated from them by tens of millions of years of evolution).

Raikova has shown that a particular developmental stage of this parasite corresponds to every stage of maturation of the host oocytes. A developmental synchronization thus exists between the host cell and its parasite. This synchronization is first seen in the fact that the oocyte appears to have two nuclei, one large and the other small. The cell then divides in two. The daughter cell with the large nucleus repeatedly divides and forms a sort of placenta that surrounds an embryonic mass arising from repeated division of the daughter cell with the small nucleus.[2] When the oocyte reaches the stage at which the yolk is formed, the parasite takes the shape of a double-barreled larva. The oocyte does not die (although its nucleus seems damaged) and continues to accumulate yolk during the lifetime of the parasite. Later, the larva transforms into a stolon—an elongated structure on which buds form, just as for many free-living coelenterates. Each bud becomes a little tentacled medusa or jellyfish. The parasite remains in the sturgeon oocyte and spends the winter in this state.[3] In the spring the parasite resumes growth by using the oocyte yolk. Finally, at spawning, the parasite is released into the water along with healthy oocytes.

The parasitic stage of this life cycle may last 10 years or more. In fact, if (as Raikova surmises) the parasite infects sturgeon while the latter are larvae, at least 10 years (16 for the beluga) pass before adulthood. Now, it is only at that moment that the transformation to the stolon form can occur, because the parasite has no means of getting outside its host, into the water, unless the female sturgeon lays its eggs.[4]

The free-living stage of this species is very ordinary: the stolon fragments into tiny, solitary medusas that eat various prey, divide, and then reproduce sexually. The cells that arise from fertilization invade the sturgeon ovaries, and the cycle begins anew. Nothing is known about how the *P. hydriforme* larvae actually get into the sturgeon ovaries (Raikova thinks there may be an intermediate host, such as a small freshwater annelid), and neither is it known if they can infest male sturgeon.

Given the current state of knowledge, it is impossible to explain why or how only *P. hydriforme* in its phylum became an intracellular parasite. Rather, this example suggests (and I will return to this point) that the passage to parasitism does not require a species to belong to a particular phylum with some specific traits. For several reasons (especially the fact that *P. hydriforme* is found both in sturgeons in Europe and in species of *Poly-*

odon in North America), Raikova believes the passage to parasitism of *Polypodium* happened during the Cretaceous period, at least 60 million years ago.

At the opposite end of the spectrum from the coelenterates and their lone parasitic species are platyhelminths, or flatworms, most species of which are parasites. The phylum contains four main classes, the turbellarians, the monogeneans, the trematodes (flukes), and the cestodes (tapeworms). The turbellarians include both free-living and parasitic species, whereas the other three classes are completely parasitic. Some flatworms, especially all the tapeworms (including the species of *Taenia*, "solitary" tapeworms),[5] are among the most difficult to relate to free-living species simply by comparing morphologies. For now, there is no definitive information on how (or how often) the flatworms underwent the passage to parasitism.

More accessible to evolutionary research are the phyla that include similar numbers of free-living and parasitic species. On the basis of the information gleaned from such phyla, it is possible to hypothesize about the passage to parasitism.

The fundamental question can be framed as follows: In a phylum of this type, did the passage to parasitism occur once or several times? If the response is once, that means that all parasitic species found today in that phylum have a common ancestor, a species that, at some moment during the evolution of the phylum, abandoned a free-living lifestyle for a symbiotic one. If the response is several times, this implies that different lineages became parasitic independently. In the latter case, a parasitic mode of life in two species need not imply a close phylogenetic relationship; it may simply reflect the phenomenon of evolutionary convergence (fig. 2.1).

The phylogeny of the nematodes has turned out to be an excellent subject for this sort of research, and molecular methods have answered many of the questions.[6] Blaxter et al. (1998) have shown that the nematode phylum comprises five major branches, and parasitic species are in each branch. The phylogeny, deduced from the divergence of nucleic acid sequences between different species, shows that parasitism arose at least four times in nematodes that use vertebrates as hosts and at least three times in nematodes that parasitize plants. These results show that nothing prevents the independent evolution of a parasitic lifestyle in different branches of the same phylum.

I should note in passing that I am not asking a question that is the inverse of the one posed earlier, that is, How many passages from a parasitic to a free-living lifestyle have arisen in a particular group of organisms?

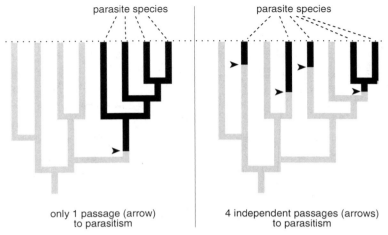

Fig. 2.1. In a given phylum, has parasitism arisen just once during the course of evolution (*left*) or several times independently (*right*)? These diagrams are hypothetical. The horizontal dotted line indicates the present time. The black arrows indicate the acquisition of a parasitic lifestyle.

This is simply because the possibility of an evolutionary return to a free-living life after acquisition of a parasitic one is a controversial matter, and it is unlikely that a single instance is currently known beyond a doubt.

An Example of the Evolution of Parasitism

A second question about the passage to parasitism is: How does a parasite differ from its free-living ancestor?

This question cannot be directly answered because it is impossible, without traveling in space–time, to observe an ancestral species and its descendents simultaneously. But we can at least compare "cousins," species believed to have a recent common ancestor, where a free-living lifestyle has been retained by one cousin and abandoned by the other in favor of a symbiotic life.

Few phyla contain a suite of species that present all the sequential stages in the evolutionary adaptation to a parasitic lifestyle. It is in groups in which most species are free-living and only several have ventured on the pathway to parasitism that we find the highest likelihood of encountering species at different stages of adaptation to a parasitic existence. Therefore, even without being able to reconstruct an authentic phylogeny directly, at least we can take a snapshot (or more precisely, a series of snapshots) showing the process of changing habitats, in the ecological sense. The prosobranch molluscs provide one of these sequences. Most species are free-living, but a series of extant species, belonging to several families, show differing degrees of adaptation to parasitism. Of course the present-day species are not derived from one another; they only allow us to imagine, by a more or less accurate reconstruction of the past, how forms and functions evolve when organisms become parasitic.

We can distinguish three stages in the reconstructed evolution.

In the first stage, the animal remains near the surface of the host, even if it has a tendency to feed more and more deeply in its tissues. There is, first of all, the "choice" of which living substrate to use, and on this substrate, the "choice" of the most favorable location. An example is the tiny prosobranch *Capulus*, which attaches itself to bivalve and gastropod molluscs, brachiopods, and even polychaete worm tubes and locates itself where a water current, generated by the "host," brings it a regular supply of prey (fig. 2.2, A). Really, *Capulus* barely merits being called a parasite, because it takes only a part of the food its host would normally ingest.

After adopting such a mode of life, the attaching species will be tempted to take its food directly from the surface or innards of the host, if the substrate is edible.[7] The substrate thus becomes a host in the strict sense of the word, and the attached symbiont becomes a parasite. It is on echinoderms, which constitute rather immobile substrates, that some prosobranch species have evolved this parasitic tendancy.[8] Species of *Thyca*, parasites of starfish, are not yet very modified. However, the ventral surface has become an attachment disk with the mouth at its center. The mouth has a powerful pharynx that serves as a suction device. The animal feeds itself exclusively at the expense of its host and has therefore become a true parasite. In certain *Thyca* species (fig. 2.2, B), the mouth is prolonged by a proboscis that allows them to draw nourishment from deeper in the host's tissues.

A more pronounced step in the evolution of parasitism among the prosobranchs is characterized by a tendency to bury itself in the host, or at least to anchor itself very firmly in the host's body. Such burying or anchoring generally begins in the integument. For example, *Asterophila* (fig. 2.2, C) anchors itself firmly in the starfish integument, which is thereby substantially deformed. This tendency toward burial is sometimes associated with the development of the proboscis, which allows the parasite to aspirate the internal body fluids of the host. The proboscis is highly developed in species of *Echineulima* and floats freely in the host body cavity (fig. 2.2, D); these parasites implant themselves in the integument of sea urchins. The *Pisolamia* have evolved a muzzle that forms a veritable rivet in the

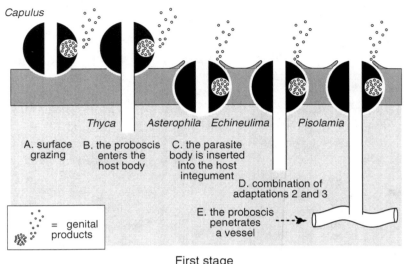

First stage

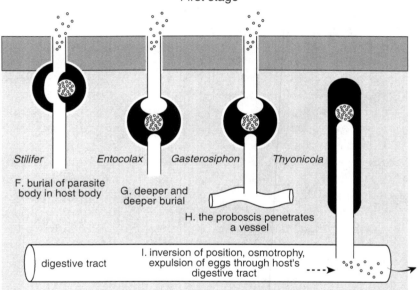

Second stage

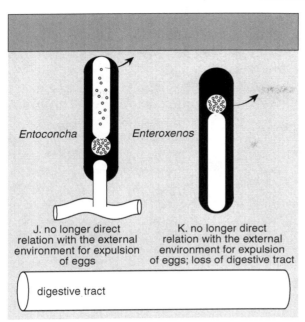

Fig. 2.2. First, second, and third stages (reconstructed from observations of extant species) in the evolutionary acquisition of a parasitic lifestyle by prosobranch molluscs. The body of the parasite, represented by a circle crossed by the digestive tract, is in black. The wall of the host is in dark gray. The host body cavity is in light gray.

Entoconcha   Enteroxenos

J. no longer direct relation with the external environment for expulsion of eggs

K. no longer direct relation with the external environment for expulsion of eggs; loss of digestive tract

digestive tract

Third stage

integument of the host sea cucumbers, and the proboscis, deeply buried, enters a vessel of the host circulatory system (fig. 2.2, E).

In the second stage, the symbiont is totally within its host but still communicates with the exterior. The invasion occurs as follows: the swimming larva of the future parasite attaches to the host integument and perforates it, after which the visceral mass grows inside the host, leaving only a thin siphon open to the exterior.

In species of the genus *Stilifer* (fig. 2.2, F), parasites of starfish, the body still remains quite close to the integument, and the proboscis floats freely in the host body cavity. In species of *Entocolax* (fig. 2.2, G), parasites of sea cucumbers, burial in the host interior is much more pronounced. Nutrition is garnered by the proboscis just as in *Stilifer*. In species of *Gasterosiphon* (fig. 2.2, H), also parasites of sea cucumbers, the siphon pierces a host circulatory vessel, just as for *Pisolamia*. In species of *Thyonicola* (fig. 2.2, I), also sea cucumber parasites, individuals enter into the host's body from the intestinal wall, not from the integument. These animals are therefore in some sense the inverse of the preceding genera.

28   Furthermore, the parasite digestive tract has disappeared and nutrition is by osmotrophy (penetration of nutrients of low molecular weight directly through the parasite's integument). Eggs are expelled through the host's digestive tract.

The third stage is the acquisition of true endoparasitism; all communication with the exterior is severed. The genera *Entoconcha* and *Enteroxenos,* both parasites of sea cucumbers, have attained this stage. Species of *Entoconcha* (fig. 2.2, J) are attached by the proboscis to a blood vessel of the host and float in its body cavity. It is thought that they enter the host through its integument and that the siphon communicating with the exterior is subsequently lost. Species of *Enteroxenos* (fig. 2.2, K) float freely in the body cavity and get their nutrients by osmotrophy (as do species of *Thyonicola*). Penetration occurs in the esophagus, to which the juvenile parasite is attached; at maturity, this attachment is broken.

Morphological Changes

In prosobranch molluscs, evolution toward endoparasitism is accompanied by a whole series of morphological and anatomical changes: disappearance of the shell, then of the twisted visceral mass itself; reduction or disappearance of most sense organs; disappearance of the anus and finally of the entire digestive tract; and formation of a pseudomantle that, in the most modified species, allows the accumulation of eggs in the form of egg-bearing capsules.

The general shape of the animal has changed greatly so that the classical mollusc silhouette finally disappears entirely: parasites buried in their hosts can have a central body prolonged at one end by the proboscis and at the other by the siphon that drains the cavity surrounded by the pseudomantle (in the genera *Gasterosiphon* and *Stilifer*). Others (*Thyonicola, Entoconcha, Enteroxenos*) are shaped like simple tubes comprising at the "oral" end the digestive tract and at the "anal" end the cavity surrounded by the pseudomantle, or just the latter cavity when the digestive tract has been lost.

The tendency toward a wormlike shape, elongated as much as possible in the prosobranch molluscs, reaches its limit in the extraordinary *Parenteroxenos dogieli,* parasite of various sea cucumbers in the genus *Cucumaria* in the region of Vladivostok. *P. dogieli,* by virtue of its elongation, is probably the "biggest" known gastropod: unrolled, it is 1.3 meters long.

All these animals, however, have larvae whose morphology leaves no doubt that they are molluscs and nearly no doubt that they are prosobranchs.

When we compare the morphology of a free-living prosobranch (fig. 2.3, *top*) with that of a parasitic prosobranch (fig. 2.3, *bottom*), we see just how far adaptation to a parasitic lifestyle can go. An interesting future avenue of research would be to compare the genome and nature of gene expression in two species that are closely related genetically but very different in appearance. One question might be, for example, How are homeotic genes (genes that determine the body plan of metazoans) expressed in *Parenteroxenos*?

Functional Changes

Apace with the progressive burial in the body of the host that I have just described, the relationships between the host and parasite and between the external environment and the parasite change radically. Consider two aspects of these relationships, nutrition and reproduction.

Nutrition is profoundly modified. From the time it becomes parasitic, a prosobranch depends entirely on its host for its subsistence, but the differ-

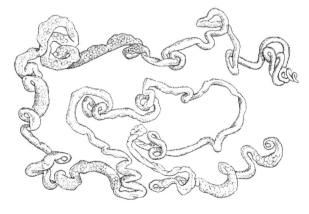

Fig. 2.3. A free-living prosobranch (*top*) and a parasitic prosobranch (*bottom*). (After Nathalie le Brun.)

ent species effect this dependency in various ways. We should note first that none of these parasites exploits the intestinal environment per se of the host. Those that are closest genealogically to their free-living ancestors consume surface tissues or at most liquids derived from digestion of these tissues outside the parasite's body. In such species, the digestive tract remains complete, with a digestive gland and an anus.

Very quickly (we are still imagining that we can follow the evolutionary process by reducing millions of years to the time needed to read these lines), "very quickly" then, the proboscis drills into the host, either penetrating the body cavity or seeking at a greater depth a circulatory vessel in which to terminate. From this point on, the digestive tract becomes simplified. If the parasite remains on the surface, it retains both a mouth and an anus. If the body of the parasite is buried in the host, the anus disappears. It is highly likely that the richness of the alimentary contents (especially when a digestive vessel has been pierced) and the low molecular weight of the metabolites that are absorbed facilitate reduction of the digestive tract (absorption pure and simple replaces digestion) and loss of the anus. The end point of this relationship of the parasite with such a rich environment is that the digestive tract disappears entirely and nutrition becomes osmotrophic.

Reproduction in all sexual organisms implies fertilization followed by dispersal of eggs, embryos, or larvae.

We would have expected the prosobranchs, which are generally gonochoristic (that is, they have separate sexes), to have evolved toward hermaphroditism as they became parasites. If we think especially of *Entoconcha* and *Enteroxenos*, prisoners in the body cavity of sea cucumbers, it seems that hermaphroditism could only be advantageous. Simultaneous hermaphroditism, which allows self-fertilization, is in effect an insurance policy against isolation because it allows an individual that finds itself alone in a host to reproduce nevertheless.

Our expectation, however, is not realized; sexes are usually separate, although male and female live near one another, often in the same microhabitat. There are, in fact, among all the parasitic prosobranchs that I have mentioned, several hermaphroditic taxa (for example, *Thyca* and *Pisolamia*), but the species that are deepest in the host body (including species of *Entoconcha*, *Thyonicola*, and *Enteroxenos*) are all gonochoristic. However, the two sexes in these species are remarkably different, and a highly original process has led to a sort of functional hermaphroditism instead of genetic hermaphroditism. In effect, animals such as I have been describing are just the females, which arise from female larvae. The male larvae inject themselves into juvenile females through the females' siphons. In some species, these males develop and remain as dwarf males in the pseudomantle of the female. In others, they attach to a specific site in the female body and lose all characteristics of individuals, becoming in effect just testicles (with a vas deferens) grafted onto tissues of the female.

As for the dispersal of eggs and larvae, the evolution of endoparasitism poses problems for molluscs as it does for all parasites. When there is still a connection to the outside, it is by this means (through the siphon) that egg-bearing capsules accumulated in the pseudomantle are expelled.

In the two strictly endoparasitic genera (*Entoconcha* and *Enteroxenos*), parasites of sea cucumbers, the solution is particularly creative. We know that some sea cucumber species eject their entire viscera, either naturally during certain seasons or when they are handled. This trait characterizes the host species of *Entoconcha* and *Enteroxenos,* and it is by this route that the parasites, stuffed with accumulated eggs, reach the outside. This fact leads to the hypothesis that this spontaneous evisceration serves the host by ridding it of the parasites or, just the opposite, that it is the parasite that provokes the evisceration to ensure dispersal of its eggs. From a quantitative view-point, it appears that the high risks of a parasitic lifestyle among the proso-branchs are compensated for by a prodigious rate of egg production.

If the mollusc example is probably the best for understanding the evo-lution of forms and functions during progressive adaptation to a parasitic lifestyle, it does not tell us much about the number of times that this pas-sage to parasitism evolved, in contrast to the nematodes mentioned earlier. It is hard to know if all the parasitic prosobranch molluscs have a recent common ancestor (that is, if the passage to parasitism evolved just once), or if a parasitic lifestyle was acquired several times in distinct lineages. The fact that the parasitic species belong to several families suggests that the lat-ter hypothesis is correct, but this subject has been little studied to date.[9]

Preadaptive Characters

With respect to the acquisition of a parasitic lifestyle, a delicate question arises: Is there a logical explanation for the very different proportions of parasitic species in different phyla? This is the question of preadaptation, or of traits that predispose a taxon toward a particular evolutionary path, and it is not an easy question.

Let us examine the three main examples I have discussed in this chapter.

First, for *Polypodium hydriforme,* which is the only one among thou-sands of coelenterate species to become parasitic on organisms totally different from itself, I have been tempted to say that this case shows that the passage to parasitism does not require any particular preadaptation in the original taxon. In fact, this enigmatic parasite misleads us. It does not provide any certain information about this matter.

Second, turning to the nematodes, we observe that the differences, at least on the morphological and anatomical levels, between the free-living species and the parasitic ones are not very great; a nonspecialist would be unable to distinguish a parasitic from a free-living nematode at a glance. From this fact, I conclude that the passage to parasitism among nematodes required relatively minor changes, so that preadaptations may have been many and important.

Third, among the prosobranchs, by contrast to the nematodes, the parasitic species have morphological traits different from those of free-living species. However, the fact that the prosobranchs are the only mollusc class in which parasitism is manifested (no opisthobranchs or pulmonates are parasites) gives us a clue. Prosobranchs are also the only molluscs that have a proboscis for sucking liquids. This proboscis, often associated with a perforating apparatus, allows them to pierce shells of other molluscs and feed on their tissues. Here is a means of predation that one can easily envision becoming part of a persistent association in which the prey becomes a host. The only condition needed for such a transformation is that the prey not be killed immediately when it is fed upon. We can then easily imagine that in attacking prey sufficiently large that they are not killed outright, the ancestral prosobranch parasite remained persistently associated with the same food source. There was thus a passage from a temporary association to a persistent one.

So it seems that particular conditions (preadaptive traits) can aid a free-living species to acquire a parasitic lifestyle. It is difficult to imagine how natural selection would operate at the very moment when the change occurs, if the free-living species did not benefit from this change. For this selective advantage to exist, certain traits of the free-living species must be adaptive at the outset of this passage, which is true for any evolutionary change in lifestyle. (For example, all the terrestrial mammals did not return to the water; we can logically conclude that those that did must have possessed, even while they were terrestrial, traits that preadapted them to an aquatic life.)

What I have just written is equally true for mutualism. D. W. Davidson and D. McKey (1993) cite the example of associations between ants and plants in which the ants inhabit particular structures (domatia) produced by the plants and in return defend these plants against phytophagous insects. Such an association could only have arisen if the ancestral plants had cavities in their stems in which ants could have sheltered.[10] We see that, contrary to the preceding case, this example illustrates a preadaptation to become a host, not a parasite.

What Changes?

The comparison of free-living prosobranchs with parasitic ones shows that morphology and anatomy can be profoundly altered when, in a phylum of free-living species, some evolve to acquire a parasitic lifestyle.

First, some organs shrink or even disappear. In the prosobranch series described earlier, *Thyonicola* and *Enteroxenos* achieve the total loss of the digestive apparatus. They garner nutrition by the passage of molecules through the integument. Another frequently cited example is that of tapeworms (the "solitary" worms of humans), which have not the slightest trace of a digestive tract. Unfortunately, in this case, even if we do not doubt that the distant ancestors of tapeworms had a functional digestive tract, it is impossible, working with present-day species, to reconstruct the intermediate evolutionary stages.

In reality, modifications observed in the acquisition of a parasitic lifestyle are only illustrations, among many possible, of the evolutionary loss of organs that cease to be useful. We know, for example, of cave-dwelling fishes that have lost their eyes, although we have evidence that their ancestors were sighted. What happened? The cave dwellers, isolated from light for many generations, lost their eyes because each time a mutation occurred that affected the eye, instead of being selected against as it would have been in a surface dweller, the mutation persisted in the cavernicolous population because it did not affect the reproductive success of the individuals bearing it. Gradually the eye lost its structure and function. Mutation pressure was stronger than selective pressure, which no longer acted on the eye. A mutation that caused the failure of construction of the eye or part of the eye during development could even be selected for, because the energy that the individual would have allocated to the ocular structure could be diverted to other functions, such as reproduction.

Parasite genes that are no longer indispensable cease to express themselves, by a process exactly analogous to the loss of the eye in cave-dwelling fishes. The genes cease to serve any function, or they even disappear from the genome when deletions (losses of DNA sequences) occur. They may even become "pseudogenes," that is, degenerate sequences that are completely unable, during the course of subsequent evolution, to retrieve their functions. This is why we know of no parasite (at least among the multicellular species) that has evolved to become free-living, even if such a passage appears to be theoretically possible.

These losses of organs have led to the frequent discussion of evolutionary "regression" in parasites. We should be wary of this term because it is not obvious why abandoning organs that have become useless in order to allocate the energy thus saved to other functions should not be perceived as progress, from the standpoint of adaptation. Although parasites give the impression of being morphologically simplified, they have only become progressively adapted to their lifestyle, and they are in fact more highly evolved than their free-living ancestors. Often they are able, during their development, to construct several organisms that differ greatly from one another. Moreover, when one compares the sizes of parasite genomes with those of free-living relatives, there is no evidence that parasite genomes have regressed.[11] I should add, finally, that regression and disappearance of organs are among the most normal processes of evolution. Blind cave fish aside, we can recall all the legless vertebrates (such as snakes and glass snakes) whose ancestors certainly had well-developed limbs. We can also compare species with lost organs to corporations that see their stock prices rise (which means that their adaptation to economic conditions has improved) when they close unprofitable factories and lay off personnel.

An interesting approach to parasitism would consist of identifying the factors that lead to different effects in different species. For example, what causes certain parasites to lose their digestive tracts, while others do not? Is it determined by which phylum they originally belonged to, which hosts they inhabit, what type of microhabitat they colonize, the antiquity of the association? This type of research has been undertaken for some groups by the use of the method termed *comparative analysis,* especially by comparing demographic parameters or life history traits.

The fundamental idea of comparative analysis is that logically the same causes should produce the same effects. A comparative analysis consists, therefore, in gathering available data for many taxa and determining if variable X covaries with variable Y. It is not a matter of expecting all the points so obtained to fall on a straight line in a plot of Y against X, but of the cloud of points being oriented in the predicted direction. In comparative analysis, the data are "corrected" for certain defects. The two most important corrections concern the sample size (data points are weighted according to the number of cases that have been studied to produce them) and phylogenetic relationship (we try to avoid using many data from closely related species; these would be redundant in some sense, because the relevant evolution of closely related species would not have been independent if it occurred before the speciation event). Just as the shape of organs is due to natural selection, so are demographic parameters such as age at reproductive maturity, mean lifespan, and fertility rate. These life history traits are selected to maximize reproductive success.

During the last few years, several researchers have studied the evolution of life history traits of parasites. The ideal approach consists of having many data, allowing a detailed comparative analysis of a taxon that contains both free-living and parasitic species. I demonstrated earlier how the passage to parasitism occurs in a lineage of free-living species, and the parasitic mollusc example showed us that in extreme cases, adaptation to a parasitic lifestyle can render morphology and anatomy unrecognizable. What about life history traits?

On this point, an interesting question arises regarding body size. It is interesting for two reasons. First is the existence of Cope's law, which says that in any evolutionary lineage, body size tends to increase. The usual example is the growth in body size of the horse and its ancestors between the Eocene epoch and the present.[12] The second reason is that there is a widely espoused viewpoint that says that parasitic species have become smaller than their free-living ancestors, thus contradicting Cope's law. It is true that, a priori, one might imagine that the limited space available within hosts constrains parasite body size. One might equally have predicted that this constraint is more pronounced in parasites of the smallest hosts (that is, one might expect the biggest hosts to have the biggest parasites, and vice versa).

The size of an organism can be measured in different ways (for example, length, volume, biomass). In the example I cite, Serge Morand et al. (1996) studied the evolution of size in a group of nematodes, the pinworms, that includes many species, some of which are free-living and others parasitic. Among the latter, some parasitize invertebrates and others vertebrates.[13] This comparative analysis supports the hypothesis that the biggest parasites are found in the biggest hosts. It shows that when a transfer occurs and the "donor" and "receptor" hosts are of different sizes, this difference generally affects the size of the nematodes.[14] A transfer to a smaller host leads to a smaller parasite, and a transfer to a larger host leads to a larger parasite. It is not the phylogenetic position of the pinworm that is crucial but the environment provided by the host. It is therefore incorrect to say that passage to a parasitic existence automatically entrains a size reduction; evolution in the other direction is possible.

36   Other research (Skorping, Read, and Keymer 1991; Morand 1996) shows that body size is correlated with various demographic traits. Positively correlated with body size are time to reproductive maturity (the bigger the parasite, the longer the juvenile period), lifespan (larger nematodes live longer), and daily fecundity (larger nematodes produce more offspring). These correlations with demographic traits differ from those in free-living animals. Among vertebrates, for example, fecundity is greatest in smaller species and lower in the largest species, just the opposite of the relationship in the nematodes.

A. F. Read and A. Skorping (1995) have proposed an interesting hypothesis about the relationship between body size and the existence of migratory movements within hosts. Certain nematodes, like species of *Ascaris*, undertake a surprising migration. Larvae of these species hatch in the host intestine after infection of the latter by eggs, but the larvae become adult only after passing through the liver, lungs, or trachea and having been swallowed a second time. Read and Skorping have observed that these migratory species are larger as adults (and therefore more fecund) than nonmigratory species. In other words, the passage through various host tissues before definitive installation in the intestinal microhabitat allows the parasite to grow more rapidly and ultimately to achieve a larger size.

As these examples show, the ways in which natural selection works in parasites after they have acquired their new mode of life are very varied. When host–parasite relationships are generally better known (which will require decades of research), it will be possible to put a general theory of adaptation to a parasitic lifestyle on a sound footing. The question How does a species become parasitic? will then be answered, and although the answers will be complex, they will elucidate one of the most remarkable adventures of life: the conquest of life itself.

Controversy: Surprising Parasitism

The foregoing sections can give the impression that parasitism is so widespread, and the word can apply to so many species, that it is difficult to see its limits. This impression is correct.

The fact that in the last twenty years of the twentieth century we became aware that symbioses, whether parasitic or mutualistic, have played and continue to play key roles in the evolution and functioning of the biosphere has led researchers to ask a new question: What is truly parasitic or mutualistic and what is not? We broach two facets of this same question that verge on the shocking: Are there parasitic molecules, and are there parasitic genomes?

Are There Parasitic Molecules?

This question Are there parasitic molecules? could not have been posed before the late 1970s. It was precisely in 1980 that L. E. Orgel and one of the discoverers of the DNA double helix, F. H. C. Crick, published an article suggesting that parasitic sequences could be embedded in the genome (Orgel and Crick 1980). At this time, researchers were questioning the significance of many regions of the genome whose sequences indicated that they could not be translated into proteins. More concisely, the question was, What are these genomic sequences doing if they are not genes? There seemed only two possible answers. Either these sequences had specific functions even if they were "noncoding" (for proteins), or they were intruders—that is, parasites.

Today, sequencing that has been completed or nearly completed for several large genomes (the plant *Arabidopsis thaliana,* the free-living nematode *Caenorhabditis elegans,* the fruit fly *Drosophila melanogaster,* humans) has convinced specialists that this question is increasingly germane. Noncoding DNA, although rare in prokaryotes (bacteria), can constitute up to 90 percent or more in genomes of certain higher organisms, including humans.

One fraction of this DNA consists of repeated sequences of which at least some part has a function in the chromosomes. These sequences are found in the centromeres and telomeres and play a role in various stages of cell division. These are not parasites. Another fraction consists of pseudogenes, sequences that one can deduce have derived from ancient genes that have lost their function in the course of evolution and in which unselected mutations have changed the original base sequence. Even if these pseudogenes are not producing anything useful and even though the cells copy them at each division, they also are not parasites. At most one might consider them as slightly bothersome leftovers, like the appendix that graces our intestine at the boundary between the ileum and the colon.

Much more interesting for parasitologists are entities called *transposable elements,* whose very name suggests they have an independent life because they can "transpose," which is to say that they can replicate outside the normal replications of the genome in which they are inserted. They owe this autonomy to the fact that they possess, in their short sequences, codes for the enzymes needed for their replication and their movement to other parts of the genome. There is hardly any doubt that they are intruders, these "selfish genes," to use the expression of Richard Dawkins (1976).

This conclusion is strengthened by two important observations. On the one hand, transposable elements are regulated by the host genomes; an

38 entire series of mechanisms acts to prevent the transposition of the elements and the expression of enzymes coded for by them from leading to a fatal invasion of the host genome. On the other hand, there exist transposable elements whose genetic structure is very close to that of retroviruses, which are unquestionably parasites.[15]

The only question that really remains to be settled, to take up the controversy introduced in chapter 1, is this: Are the transposable elements parasites or mutualists? We know the response rests on the entire balance sheet of costs and benefits that they confer on the genome that carries them. The cost arises from the fact that they consume molecular material (purine and pyrimidine bases, sugars, phosphoric acid) needed to construct any DNA molecule. The benefit, long ignored, is that the insertion of these elements provokes mutations that are a priori neither good nor bad but that contribute to the continuing generation of genetic diversity. It is very difficult given the current state of knowledge to say if the benefit exceeds the cost or vice versa. But there is a tendency among many biologists sensitized to the importance of genetic diversity to assume that the benefit outweighs the cost.[16] Parasitologists are firmly in this camp, because it is almost certain that transposable elements have, for example, played a role in the evolution of variability in the immune system of vertebrates. They have even played a role in all of evolution by favoring the reproductive isolation of species through the shuffling of chromosomes that they can induce (Ladevèze et al. 1998).

Are There Parasitic Genomes?

We have long known that there are parasitic chromosomes. These are called B chromosomes, in contrast to normal A chromosomes. B chromosomes are selfishly transmitted, in ways similar to the ways transposable elements move. They have been described from hundreds of species of animals and plants, but the best known are the B chromosomes of the parasitic wasp *Nasonia vitripennis*. As in all Hymenoptera (ants, bees, and wasps), fertilized eggs (diploid) produce females, whereas unfertilized eggs (haploid) produce males. The sperm (and not the egg) can carry the B chromosome. When it does, the B chromosome causes the total disappearance of the rest of the paternal genome so that the egg, although fertilized, becomes a haploid male. The B chromosome thereby assures its own transmission to the next generation, because it is present only in the sperm. The parasitic status of B chromosomes is certain, because, even if their pathogenic effect is manifested only at the moment of reproduction, they persist in the cells they infest from the moment those cells are formed.

Only recently have we begun to suspect that entire genomes can behave as parasites. Here are the facts, summarized for the most part from research by Alain Pagano (1999).

Several species of green frogs exist in Europe. To simplify matters, I focus on only two of them, *Rana ridibunda* and *Rana lessonae*. How do these differ? Not by much, but there are differences: *R. ridibunda* is larger than *R. lessonae; R. ridibunda* prefers running water, while *R. lessonae* is usually found at the edge of stagnant water bodies. Biochemically they are distinct; all tests, including analyses of DNA, distinguish the two species unambiguously.

Although distinct, however, the two species are evolutionarily close, which allows them to hybridize. But not in any old way. Male *R. lessonae* easily mount female *R. ridibunda*, whereas male *R. ridibunda* are too big to mount female *R. lessonae*. The only possible hybridization is therefore male *lessonae* × female *ridibunda*.

We would expect logically that, at meiosis, the gametes of the hybrid progeny would have equal numbers of *ridibunda* chromosomes and *lessonae* chromosomes, and even that there would be crossovers (recombinations) between the maternal and paternal chromosomes of a chromosome pair. However, nothing of the sort happens, at least as far as we know. In all gametes of the hybrids, we find only genetic material of *ridibunda* origin. All genetic information that came from the *lessonae* parent has been eliminated. This process can repeat itself indefinitely from generation to generation (fig. 2.4).

If we analyze closely this astounding mode of interaction between the two species, we notice first that the association of information derived from *ridibunda* with information derived from *lessonae* is such that, during the entire life of the individual, the two genomes collaborate with one another, first to construct the body of the animal that contains them and then to make it function. The proof of this statement is that the hybrids manifest many traits that are intermediate between those of the two parents.

But everything changes at meiosis. The apparent cooperation between the two genomes suddenly ends, and the *lessonae* genes are not transmitted to descendants.[17] This is why Pagano calls this interaction parasitism, although it is very different from a classic parasitic association. The question that immediately arises is this: What advantage does the *ridibunda* genome derive from this manipulation? The response is not obvious, but Pagano's observations cause us to look toward the ecology of the two species. It seems that this unusual parasitism allows the *ridibunda* genome to exploit environments (stagnant water) in which the pure *ridibunda*

Fig. 2.4. Does the genome of the green frog *Rana ridibunda* act as a parasite of the genome of another green frog, *Rana lessonae*? Frogs labeled RR are "pure" *ridibunda*, those labeled LL are "pure" *lessonae*, and those labeled RL are hybrids, *ridibunda* × *lessonae*.

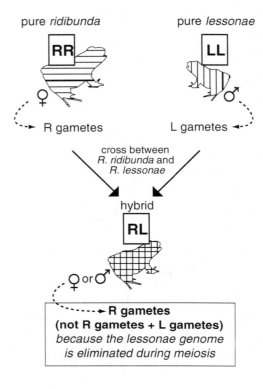

pure *ridibunda*

pure *lessonae*

RR

LL

♀ R gametes

L gametes ♂

cross between
R. ridibunda and
R. lessonae

hybrid

RL

♀ or ♂

**R gametes
(not R gametes + L gametes)**
*because the lessonae genome
is eliminated during meiosis*

phenotypes would not have been competitive with *lessonae* phenotypes. In fact, hybrids are frequently found in the same pools as pure *R. lessonae* individuals.

The unusual parasites I have just discussed provide strong arguments for biologists who see in DNA molecules the real actors of biological evolution, actors for whom their phenotypes are but tools (cf. Dawkins 1976).

The Profession of Parasite

3

Parasites Aim at Living Targets, the Hosts

Every living organism has a life cycle, if only because it has a beginning (birth) and an ending (death).[1] The life cycle includes morphological and physiological changes from the first stages of development until maturity and old age. The dandelion seed attached to its umbrella and carried by the wind is nonetheless a dandelion, though a dandelion very different from an adult plant. These changes become especially dramatic when a species exploits several different environments during its life cycle. We then see real metamorphoses. For example, many insects and amphibians must live first in an aquatic habitat and then in a terrestrial one. Everyone knows that a dragonfly nymph is as profoundly different from an adult dragonfly as a tadpole is from a frog.

In parasites and mutualists, there are always major metamorphoses, even series of metamorphoses. The reason is that the parasite or mutualist is confronted with extreme environmental changes in the course of its life cycle. With very rare exceptions, there exist at least two distinct habitats in a parasite's life: the one it occupies in or on its host and the one it occupies during a transition between two successive individual hosts. Just the fact that the individual host is mortal means a parasite needs a path to leave the dying or dead host and find another living host.

The transitional habitat can be simply the external environment. For instance, intestinal amoebas (*Entamoeba histolytica*) produce cysts that are expelled to the external environment, where their thick wall

An unoccupied ecological niche, an unexploited opportunity for living, is a challenge.

◊

T. DOBZHANSKY (1973)

41

42 allows them to resist desiccation. It suffices for them to be ingested, for example, with vegetables or contaminated water, to initiate a new infection. The transitional habitat can be a living organism of a different species from the first one. For example, the larvae (filariae) of the nematode *Wuchereria bancrofti* that causes the disease elephantiasis live in human lymph vessels, where they cause the spectacular, even monstrous, swelling of the legs and scrotum that characterizes the malady. After copulating, females expel huge numbers of minuscule larvae, the microfilariae, which spread throughout the bloodstream. To continue their development, these microfilariae must be ingested by a mosquito, in which they undergo further development leading to an infective stage. Finally, the mosquito in the course of a blood meal will be able to infect another human; the microfilariae fall on the skin and penetrate the human host through the mosquito bite.

In many instances the transitional pathway can be much more complex. As an adult, the trematode *Halipegus ovocaudatus* lives under the tongue of green frogs. There are also (1) a first aquatic larval form, the miracidium; (2) a parasitic stage, the sporocyst (or redia), which parasitizes molluscs; (3) a second aquatic form, the cercaria; (4) a stage that parasitizes copepods (crustaceans), the mesocercaria; and (5) a stage that parasitizes dragonfly nymphs, the metacercaria. When a frog eats the adult dragonfly, the ingested metacercaria transforms to an adult and the *Halipegus* life cycle starts anew.

The complexity of parasitic life cycles cannot fail to challenge the biologist. From the perspective of gene expression, the more complex the cycle appears, indeed to the point of becoming disconcerting, the more exciting questions it poses. In fact, depending on the suite of physical or chemical signals gleaned from the environment, a parasite or mutualist must in effect be able to cause the expression or suppression of entire fractions of its genome. The questions that arise include ones such as the following: Which signals are used? Where are the receptors of these signals? Which genes act to switch other genes on and off? From an evolutionary perspective, we can ask still other questions, just as difficult to answer. For example: How did these life cycles evolve? In which order were hosts added as a multihost life cycle evolved? Do life cycles tend evolutionarily to become simpler or more complex?

The helminths (that is, all the platyhelminths [flatworms], nematodes [roundworms] and less numerous worms like the acanthocephalans [spiny-headed worms]) provide the best possible example of a diversity of transmission modes. Figure 3.1 summarizes the main variants of these life cycles while taking into account only three aspects: the alternation between

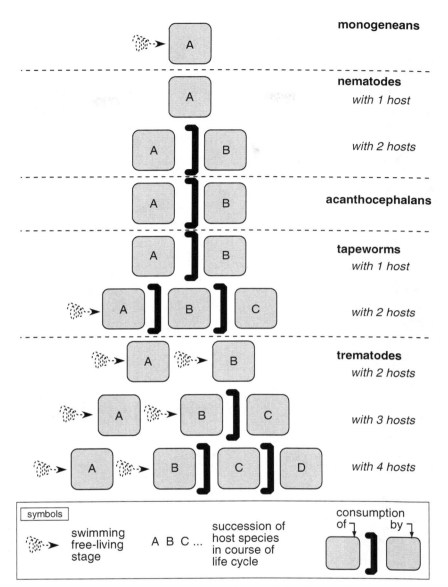

Fig. 3.1. Free-living phases and parasitic phases in the helminth life cycle. Each host species is symbolized by a square with rounded corners.

44 life in a parasitic state and passage through the external environment; the possible existence of a free-living aquatic stage (indicated in the figure by a spiral ending in an arrow); and the possible existence of a process in which a host consumes another host (indicated by a backward C).

What determines whether one life cycle, among many in the same parasite taxon, is more or less complex—for example, by comprising two hosts, or three, or more? Figure 3.1 gives the impression of great complexity. We see all sorts of different means of reaching the same goal: to return to the beginning of the life cycle. In another book (Combes 2001) I have ironically titled the chapter devoted to parasitic life cycles "Why Make It Simple When You Can Make It Complex?" In fact, we are entitled to ask why certain life cycles entail so many changes of habitat. We sometimes have the impression of a business traveler who goes from New York to Washington via Bermuda.

The truth of the matter is that, for parasites, the route via Bermuda is sometimes more profitable. We can demonstrate this fact by comparing two closely related tapeworm species, both parasites of marine fishes. These two tapeworms are *Bothriocephalus gregarius,* found in turbot, and *B. barbatus,* found in brill. Turbot and brill are both large flatfish and are morphologically similar; they coexist in the same environments in the Mediterranean and the Atlantic. Several field studies show that whether measured in terms of prevalence (fraction of fishes parasitized) or intensity (number of parasites per fish), *B. gregarius* is always more common in turbot than *B. barbatus* is in brill. In other words, it seems as if *B. gregarius* is more "successful" than *B. barbatus.*

Do the life cycles differ? Yes. Does a simpler life cycle explain the greater success of *B. gregarius?* Are you willing to guess yes? Well, you are mistaken. It is exactly the opposite; *B. gregarius* is more successful because its life cycle is more complicated. I should explain. *B. barbatus,* the less successful parasite, has a two-host life cycle. Eggs give rise to a swimming larva (the coracidium), which a copepod must swallow if development is to proceed. In the copepod, the coracidium changes into an infective stage (the plerocercoid). When a brill swallows a parasitized copepod, the plerocercoid becomes an adult parasite.

B. gregarius, the more successful parasite, has a similar basic life cycle, but it adds a detour. If a small fish such as a goby eats the copepod, the plerocercoid survives in the fish digestive tract, and the turbot becomes infected when it eats gobies, rather than eating copepods directly.

Detailed studies of the ecology of this life cycle, plus mathematical modeling (Morand, Robert, and Connors 1995), have shown that the detour through the goby increases the probability of infecting the turbot. In fact, (1) turbot, except when they are very young, rarely eat copepods, which are too small to satisfy their appetites; (2) gobies, by contrast, are only a few centimeters long and eat many copepods; and (3) turbot eat many gobies.

These facts demonstrate that the passage from a two-host life cycle (copepod–brill) to a three-host life cycle (copepod–goby–turbot) increases the parasite's success rate. The addition of a stage to the life cycle, far from reducing the probability of successful completion, enhances it. We should not, of course, consider complex life cycles as a sort of feat that parasites deliberately set out to accomplish. They are nothing more than the results of natural selection.

It is important to know that a molecular study of plerocercoids found in naturally infected gobies shows they all were *Bothriocephalus gregarius,* the parasite of turbot. We can obviously ask why natural selection has not led to the addition of another host to the life cycle of *B. barbatus,* the parasite of brill. We can only hypothesize. We might imagine, for example, that either the mutations needed for the detour through the goby have not arisen in the *B. barbatus* genome or that the association between *B. barbatus* and brill is too recent for natural selection to have had time to operate.

From the preceding example, however, we should not conclude that natural selection always leads to complexity. Figure 3.2 details one group of the life cycles depicted in figure 3.1, those of trematodes with two or three hosts. These cases show that simplification can also be selected. Represented first in figure 3.2 is what we call the "classic" trematode life cycle, comprising three successive hosts.

1. A molluscan host always begins the cycle. This mollusc is generally penetrated by a swimming larva, the miracidium, which, as soon as it has entered the mollusc, transforms into a sort of sac, the sporocyst. Inside the sporocyst are found "germ cells" that multiply to form either other sporocysts (called secondary sporocysts) or more complex organisms (which have a pharynx), the rediae.[2] Inside the secondary sporocysts, or the rediae, the cercariae differentiate, and these typically escape from the mollusc and swim in open water (in chap. 8, devoted to how parasites occupy space, we will see that this escape to the outside does not always occur). Figure 3.3 depicts the astonishing metamorphosis of a "typical" trematode, *Nephrotrema truncatum,* a parasite of shrews.

2. The cercariae enter a vector in which they encyst (generally) in the form of metacercariae without multiplying or undergoing striking morphological changes. This vector can belong to any of several different taxonomic groups.

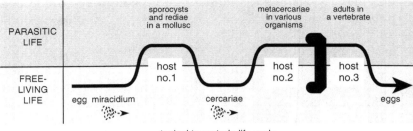

typical trematode life cycle

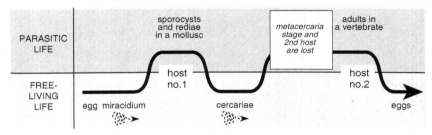

trematode life cycle, type *Schistosoma mansoni*

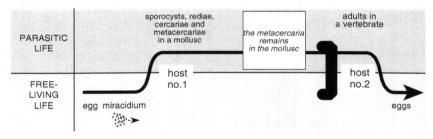

trematode life cycle, type *Microphallus pygmaeus*

Fig. 3.2. Typical trematode life cycle (*top*), reduction to two hosts in *Schistosoma* (*center*), and reduction to two hosts in *Microphallus* (*bottom*). The backward C indicates that the species to the right eats the species to the left (see fig. 3.1).

3. When a vertebrate eats vectors of metacercariae, the parasites settle either in the digestive tract of the new host or in associated organs such as the liver or lungs, becoming hermaphroditic adults and laying eggs. These are expelled from the host, and the cycle begins anew. Some species leave the digestive tract and achieve unexpected migrations. For example, when shrews ingest the metacercariae of *Nephrotrema truncatum* (the vectors are freshwa-

ter annelids), the juvenile parasites puncture the intestinal wall, detour through the liver, then travel through tissues to the right kidney (never the left!), finally settling in the renal pelvis (Jourdane 1974).

Figure 3.2 next shows two cycles, those of *Schistosoma mansoni* and *Microphallus pygmaeus*, in which the number of hosts is reduced to two.[3] How can we explain this difference? Of course, any explanation can only be speculative because no one actually witnessed the changes that occurred in

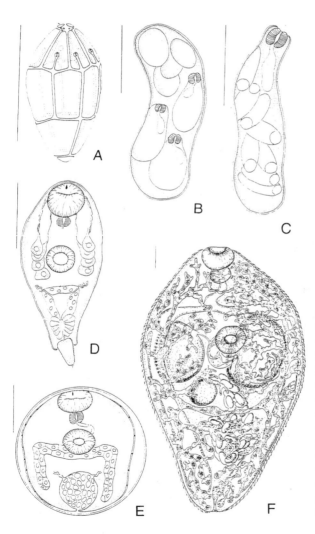

Fig. 3.3. Typical trematode life cycle, exemplified by *Nephrotrema truncatum*, parasite in kidneys of the European shrew *Neomys fodiens*. (A) Miracidium (cilia not depicted); (B) sporocyst; (C) redia (note pharynx); (D) cercaria (note two suckers); (E) encysted metacercaria; (F) adult. Lines at left of each stage represent 100 micrometers.

48   the course of trematode evolution. It nevertheless seems logical to reason as follows.

For the *Schistosoma mansoni* life cycle we observe that the cercaria penetrates a vertebrate, which will be a crocodile, a bird, or a mammal. Is a typical three-host life cycle possible in these circumstances? For this pathway to exist, the host the cercariae have entered would have to be eaten by a third host. This would be possible in some cases (bird or small mammal) but difficult in all the others because the crocodile, most birds, and all the large mammals (there are schistosomes in elephants) are "terminal" consumers that either are not eaten by anything or are eaten only after years or decades. We can thus conclude that the two-host life cycle of schistosomes is an adaptation to penetration by the cercariae of rarely eaten hosts. The cercaria, instead of transforming to a metacercaria, becomes an adult directly.

The solution to this enigma differs for the *Microphallus pygmaeus* life cycle. All *Microphallus* species as adults parasitize seabirds, such as gulls. Now, such birds normally consume huge quantities of small gastropods that inhabit beaches or rocks exposed at low tide, and these gastropods are the first hosts of *Microphallus*. It is thus surely advantageous for *Microphallus* to live and reproduce in the mollusc without incurring the risks associated with dispersing in seawater. We will see later (chap. 8) that this type of cycle confers two other advantages. One is that it allows infections en masse (the bird swallows in one gulp hundreds of infective stage individuals). The other is that it allows colonization of very harsh environments, such as polar seas (the cercaria would not directly confront cold, turbulent water). It therefore seems reasonable to believe the reduction to two hosts resulted from natural selection.[4]

The preceding comparisons show that parasite life cycles should be interpreted as adaptations among many other adaptations, just like morphological, physiological, and biochemical traits. Simplicity or complexity is retained by natural selection, the choice determined by whichever is more advantageous, with intermediate degrees of complexity possible. This is not to say that life cycles such as those I have discussed constitute the best possible adaptation to current conditions (the slow pace of evolution means that there is often a lag between current conditions and adaptation).[5]

It is often said that evolution is opportunistic and cobbles things together. François Jacob popularized the term *tinkering* in a 1977 article showing that one of the most important constraints on evolution is that it does not really make things anew but rather modifies things that already exist. This is true at the level of genomes: "new" genes can arise only by modification or duplication of preexisting genes. It is just as true at the phenotype level: "new" structures are constructed from old ones.

When a life cycle includes several host species in succession, the parasite finds itself confronted, as just described, by living environments that may differ greatly. The trematode *Schistosoma mansoni*, for instance, as we have seen, exploits a mollusc as a larva and a vertebrate as an adult. The vertebrate can be a human or, in some circumstances, a black rat (*Rattus rattus*). Now, a mollusc and a vertebrate differ not only in their morphology and immune system but also in their life histories. By life history I mean, for example, developmental rate, longevity, maximum size, and mobility. The fact that it exploits two successive hosts with different life history features implies that the parasite is adapted to these differences. A noteworthy result is different characteristics of the infections of the different hosts, in both quantitative (demographic) and qualitative (genetic) ways. Few detailed studies explore this aspect of parasite adaptation; here I summarize some aspects from studies by Théron et al. (2004) on *Schistosoma mansoni*.

For several years, these scientists have undertaken molecular ecology research in a part of the West Indies where black rats harbor adult schistosomes. Using DNA analysis, they have established two main results:

1. With respect to demography, the infection rate (prevalence) averages 0.6 percent among molluscs but 94 percent among rats; the vertebrate host population confronts *S. mansoni* much more frequently than does the mollusc population.

2. With respect to genetic diversity, an infected mollusc contains on average only 1.1 different schistosome genotypes, whereas each rat is on average infected by 34 different genotypes.[6] This indicates that each infected mollusc, with rare exceptions, contains just one miracidium, whereas the average infected rat contains many genetically distinct cercariae. (We can deduce from these statistics that infected humans contain at least as diverse a schistosome complement.)

These remarkable results show to what extent exploitation of different hosts (in this case, even by the same parasite individual) leads to differences in the nature of the infection, depending on differences in the host life histories. Morphological, anatomical, and biochemical differences in a single parasite species at different steps in its life cycle are well known. The research by Théron et al. sheds new light on differences in the nature of parasitic infections, linked to host life histories. It illustrates the great diversity of adaptations to a parasitic lifestyle. We can assume that the adaptations in the schistosome case have been selected simultaneously in the intermediate host (the mollusc) and the "definitive" host (the vertebrate),

50 so this parasitic life cycle shows that evolution can simultaneously engage in two different tinkerings (to use Jacob's metaphor).

A Parasite Should Encounter Its Host

The Parasite Picks up Information That Comes from the Host

In the biosphere, all interactions are matters of circulation and reception of information.[7] In predator–prey systems, each "partner" in the system emits information (odors, noises, movements) that can be gleaned by the other. Cats and mice do all they can not to emit too many signals, while each attempts to detect signals from the other. Propagules (elements produced by reproduction, such as eggs, seeds, or spores) survive only if they find a suitable environment. Often such an environment is immediately available: trout eggs hatch in freshwater; young pandas are born in bamboo forests. If dispersal occurs in such cases, it is to avoid intraspecific competition, and a risk is that in moving they will not find a suitable environment.

When we turn to immobile species (plants or sedentary animals), the question is posed in less optimistic terms. The dandelion seed carried by the wind need not fall on suitable soil; the swimming oyster larva or that of the fan worm need not find conditions propitious for its survival. In such species, the low probability that the propagules will end up in a favorable habitat is generally compensated for by greatly increased fecundity. Wastage is inscribed in the genes, so to speak.

With parasites, the difficulty is even greater for two reasons. The first is that parasites are typically very host-specific (see chap. 8); this means the living environment in which they can survive is highly circumscribed. If the dandelion seed can perhaps germinate in various kinds of soils, the cysticercus of human tapeworms can develop only in humans. The second reason is that the hosts in which a parasite can live constitute a fragmented habitat, dispersed and mobile, divided into as many fragments as there are host individuals. We often compare a population of hosts to an archipelago. Anything that is not a host is a lethal environment for parasite propagules. These are analogous to shipwrecked humans in a hostile ocean, desperately seeking islands that are never very numerous.

We are not surprised that natural selection has provided parasites on the one hand with fecundity rates similar to those of sedentary animals (Louis Euzet defines parasites as sedentary animals on mobile substrates) and on the other hand with strategies that allow them to maximize the probability of landing on one of the life-saving island hosts.

A fact that ought to favor the parasites is that the island (the host individual) carelessly disseminates information on its existence in the form of various signals (visual, olfactory, or acoustic). In principle, then, the parasite can detect the signal and respond by nearing the island. These signals emitted by hosts play a role in the transmission of parasites that depends on the mode of infection (Combes, Bartoli, and Théron 2002).

Figure 3.4 shows that there exist four major categories of parasites that are distinguished by how they attach to or penetrate their hosts. Parasites indicated as type A are content to attach to the host integument; these are ectoparasites.[8] Type B parasites actively penetrate through the integument; they become endoparasites. Parasites of type C use biting organisms (usually arthropods) to inject them through the host integument; like type B species, they become endoparasites. Type D parasites are passive in that they are ingested by the host, often with the host's food. They become mesoparasites (some of them, however, eventually puncture the wall of the digestive tract to become endoparasites).

A consequence of this diversity of invasion modes is that the mechanisms of detection and response and their efficiency differ greatly from species to species. Let us quickly put aside parasites of types C and D. By using vectors, type C parasites can dispense with any investment in signal detection. The species of *Plasmodium* carried by mosquitoes, *Leishmania* carried by phlebotomid flies, *Onchocerca* transmitted by blackflies—all these parasites have the greatest confidence that their insect vectors will find your skin. It is these insects that detect the host signals and respond to

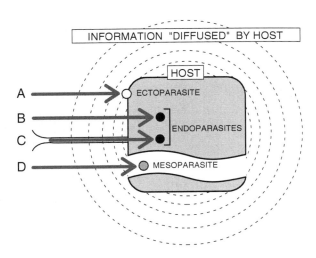

Fig. 3.4. The four major modes of parasite invasion. Type A and B parasites use information provided by hosts to find and identify them. Type C parasites, carried by a vector, leave to that vector the task of using information from the host. Type D parasites, passively ingested by the host, do not use information provided by the latter.

them, not the parasites themselves. As for type D parasites, they are the laziest of all because they detect nothing whatsoever but wait around until a potential host comes to them. The amoebic cyst, the metacercaria of the liver fluke, and the tapeworm cysticercus are able to wait for weeks or months until a stroke of good fortune brings them to your plate. This does not keep them from using strategies other than reliance on providence, as we will soon see.

It is only in parasites of types A and B that any ability of the infective stage to detect signals emitted by a host becomes adaptive and therefore confers a selective advantage. Are such parasites adept at doing this? Not really. Visual signals seem not to be used; no parasite "sees" its host.[9] Usually it is a chemical signal (odor or chemical gradient) that is detected.[10]

For instance, this is the method used by the trematode miracidium that has to penetrate a mollusc, the wasp that lays her eggs in figs, and insect parasitoids that lay their eggs in caterpillars. In this latter case, initial detection is often of the odor of the plant the caterpillar inhabits, then subsequently, at a shorter distance, the odor of the caterpillar itself. Some parasitoids can even distinguish by olfactory means plants damaged by caterpillars from undamaged ones. Detection of vibrations is rarer, but I should mention hymenopteran parasitoids that can detect beetle larvae through several centimeters of wood and reach them with their ovipositors.[11] Research using very refined acoustic detection methods (Meyhöfer, Casas, and Dorn 1997; Meyhöfer and Casas 1999) shows that in parasitoid–host systems, the aggressor uses acoustic information to detect its host and the potential victim uses such information to detect the aggressor, just as in some classic predator–prey systems. Worth noting is the astonishing mite *Histiostoma laboratorium,* which can detect the beating wings of an approaching fruit fly and jump onto it in full flight.[12]

Detecting a signal is advantageous only if it is possible to respond to it appropriately. In fact, few parasites respond in a truly effective way to signals. I observed, for example, that trematode miracidia are sensitive to molecules emitted by molluscs, but it is important to add that their ability to orient toward the mollusc does not generally come into play except at very short distances (rarely more than a few millimeters).

Insect parasitoids and ticks, however, can be far more efficient. In insect parasitoids, it is the adult, richly endowed with sensory equipment, that seeks a host for its offspring. The behavioral response to signals emitted by the host is often remarkably effective. In ticks, there is an extraordinary organ, Haller's organ, that seems to have been "designed" with the same care as the most sophisticated detection systems of modern armies. This organ has at least fifteen microdetectors, some of which are sensitive to heat, others to vibrations, yet others to odors. The tick larvae and nymphs await their hosts (various mammal species) on twigs or blades of herbs. It is not surprising that a hiker in shorts and his dog when they pass through underbrush are both favored targets of ticks.

Upon contacting a host, parasites that actively penetrate a host can generally identify their target. However, this identification is not always as precise as might be expected. If bathers sometimes emerge very red from swimming in lakes, it is because many cercariae of schistosomes that normally parasitize ducks have erroneously passed through their skin. To mistake a human for a duck does not indicate a highly refined discriminatory sense.

Thus, to "capture" their hosts, parasites, in their infective stages, are far from having the effective adaptations that predators have to capture their prey. The cheetah is able to smell a gazelle from far away, to distinguish it among savanna plants from hundreds of meters, and to pursue it with prodigious speed. Parasites have no such powers: a cercaria barely smells anything, it sees nothing, and it swims ludicrously slowly. Natural selection compensates for this relative weakness in the response to signals emitted by hosts with new types of adaptations. By "knowing" the life cycle of their hosts, infective stages of parasites do not "know" where the hosts themselves actually are, but they "know" where they ought to be.

The Parasite Picks up Information That Comes from the Host's Environment

I have shown that parasite life cycles are often highly complex. This complexity implies that the parasite travels through the ecosystem to go from the host it is leaving (usually as a propagule) to the host in which it will continue its development.[13] Whether the latter is the same species as the host just left or a different species, it is in the best interest of the parasite to maximize its probability of encountering a new host.

When I ponder this problem, I recall what a friend who hunts woodcocks told me. Woodcock is a heralded dish, perhaps the king of game birds, but it is as difficult to hunt as it is delicious to eat.[14] The hunter therefore pays great attention to two things: being present exactly where and when he ought to be. This forces him to learn by experience which environments woodcocks favor (which type of valley floor, which vegetation type, which tree species) and the precise moments, during the day and during the year, when woodcocks fly (above all in the spring and very early in the morning).

This summary description of the behavior of woodcock hunters readily translates into how a parasite should hunt its host. There are but two differences. One is, of course, that the hunter uses his intelligence and his memory, whereas the parasite uses only genetically determined behavior. The other is that the hunter can return empty-handed without any great consequence other than the disappointment of his Sunday guests, whereas the parasite must absolutely find its host in order to survive. Like the hunter, a parasite must therefore rendezvous with its "game" in space and time. Let us first consider how parasites circulate in space.

The limited ability of parasite infective stages to perceive information diffused by potential hosts is often compensated for by the localization of the parasite in parts of the environment where the host is most likely to be. This is a true rendezvous in space. Of course the success of the meeting is far from guaranteed, but at least the parasite will have a better chance by being in the favored location of the target species, just as the woodcock hunter will.

The Rendezvous in Space

Let us approach a lagoon in Provence, warm and calm under the Mediterranean sky: wet banks, some surface ripples, a few fish barely glimpsed, while plovers, gulls, and formally attired terns fly about. In the two or three meters of water, every day in the prime season, millions and millions (perhaps millions of millions) of parasites have urgent meetings with their hosts. Consider trematodes as an example. Every trematode begins its life in a mollusc, from which emerge the cercariae produced by an intense multiplicative process. An infected mollusc can produce cercariae for months (sometimes for years!) and emit between several dozen and several thousand cercariae each day. The hunt begins.

I have chosen four trematode species: *Meiogymnophallus fossarum, Cardiocephalus longicollis, Nephromonorca lari,* and *Maritrema misenensis.* In the life cycles of these trematodes, the molluscs that constitute the first hosts (bivalves for *M. fossarum,* prosobranch gastropods for the three others) rest on the bottom of the lagoon; the vertebrates that are the final hosts are all seabirds, such as gulls. Between the molluscs and the birds is an intermediate host, and it is this intermediate host that the cercariae penetrate. This intermediate host carries the parasite to the bird by becoming the bird's prey.

The cercariae, emerging on the bottom of the lagoon and living but a few hours, thus have a rendezvous with this intermediate host. Figure 3.5, based on pioneering research of Pierre Bartoli (see Bartoli and Combes

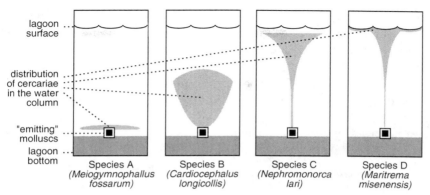

lagoon surface

distribution of cercariae in the water column

"emitting" molluscs

lagoon bottom

Species A	Species B	Species C	Species D
(Meiogymnophallus	(Cardiocephalus	(Nephromonorca	(Maritrema
fossarum)	longicollis)	lari)	misenensis)

Fig. 3.5. How the infective stages (cercariae) of four trematodes partition the water of a lagoon according to the type of hosts they infect. (After Bartoli and Combes 1986, modified.)

1986; Combes 1995), shows that the cercariae station themselves in the water column in four very different ways. We should be interested in the route leading from the mollusc, where the cercariae are produced, to the intermediate host that the gull or other bird will eat.

This route can be quite simple. This is the case when the host that is going to carry the metacercariae lives in the immediate vicinity of the mollusc that "produces" the cercariae. The first and second hosts can then both be bivalves that, even though they belong to different species, live side by side in the sediment. The cercariae of *Meiogymnophallus fossarum*, emitted by the first mollusc (*Scrobicularia plana*), are simply aspirated by the inhalant branchial currents of the second mollusc (various species) and encyst in its tissues.

The route that leads to the intermediate host can be longer. Although the molluscs serving as initial hosts are all benthic species, the intermediate host is often a fish that lives in open water or even near the surface. In such cases, the cercariae are always good swimmers who escape from the mollusc and are attracted toward the surface. Cercariae of *Cardiocephalus longicollis* ascend toward the surface, then cease swimming for awhile as they slowly descend again, sweeping through the middle layers of water where fishes of the families Sparidae and Belonidae are found; they encyst in these. Cercariae of *Nephromonorca lari* form metacercariae in fish of the family Atherinidae, which inhabit the uppermost layers of the lagoon. It is not surprising then that these cercariae climb still higher toward the surface.[15]

Cercariae of *Maritrema misenensis* travel the longest route. The initial host is a benthic mollusc as for the previous species, but the intermediate host—the target of cercariae—is a small crustacean that lives on the beach at the edge of the lagoon, the amphipod *Orchestia mediterranea*. To reach this host, tied to lagoons but not permanently living in water, cercariae of *M. misenensis* attach by their oral suckers to the very surface of the water and "wait," carried on currents, to be thrown on the shore by sea spray or waves. But this manifestly complicated route is no obstacle to completing the life cycle: Prévot, Bartoli, and Deblock (1976) show that the proportion of *Orchestia* parasitized can reach 100 percent and that a single individual can harbor up to 200 metacercariae.

This sort of study points up the degree to which strategies to rendezvous in space are essential to completing the life cycle. Of course, very many cercariae are produced, but natural selection is not relying solely on the great number. The genetically determined behavior of cercariae is such that, in each life cycle and for each host that is to be parasitized, the probability of reaching that target is maximized.

When one examines any of the trematode life cycles just discussed, it is not readily apparent that a genetic trait, despite enormous losses of individuals, has successfully traversed the difficult path leading from a living mollusc on the bottom of the sea to the digestive tract of a gull. Just as a radio signal weakens with distance but can nevertheless be captured by the appropriate receiver, so the genetic information of each parasite, after heavy loss of life, reaches the target host in which an amplifier (sexual reproduction) restores the strength of the signal.

The Rendezvous in Time

Let us abandon Mediterranean lagoons for an unattractive pond in tropical savanna. The water is more or less muddy, the insects numerous. Nevertheless, water is in short supply and necessary for life. Early in the morning, various wild or domesticated ungulates have come to quench their thirst, lazily soaking their hooves in the water. Later, toward the end of the morning, women arrive from the nearby village, washing some utensils, surrounded by children who have waded in the water. Finally, in the evening, at nightfall, rodents run all around the pond and do not hesitate to swim to capture some prey item.

The parasites I am going to place in this scene are trematodes, as in the lagoon, but their life cycle differs. They belong to the second kind of trematode I have mentioned, whose cercariae penetrate vertebrates directly. These are the schistosomes. I will describe their life cycle in some detail because schistosomes cause one of the worst parasitic diseases of the tropics—schistosomiasis or bilharziasis. The life cycle is relatively simple (two hosts instead of the three in the preceding examples): the cercariae emerge from molluscs and swim in open water, where they live several hours and are able to adhere to a suitable vertebrate host when they contact one, penetrating the skin in several minutes. To swim, the schistosome cercariae, which measure one-third of a millimeter, have a forked, vibrating tail. They shed it when they pass through the skin, and the remainder of the cercaria then undertakes a complicated migration in the circulatory system of the vertebrate host. After three weeks, the parasites reach the liver; there they differentiate into males and females, which copulate and reach the blood vessels near the digestive tract. There the females lay many eggs, which cross the intestinal wall,[16] are evacuated to the outside, and give rise to the miracidia that penetrate molluscs. The cycle begins anew.

Let us return to the African savanna, by the pond. This is the transmission site for the schistosome species *Schistosoma bovis, S. haematobium,* and *S. rodhaini.* André Théron (Théron 1984; Théron and Combes 1988) has analyzed the rhythm of many trematode cercariae as they emerge from molluscs and shown these rhythms to be genetically determined. In our African pond, cercariae of *S. bovis* (host: ungulates) emerge from molluscs early in the morning, and emergence ceases toward mid-morning. Cercariae of *S. haematobium* (host: humans) emerge essentially in the middle of the day, whereas cercariae of *S. rodhaini* (host: rodents) appear just after nightfall. The rendezvous strategy is just as evident as in the lagoon, but here it is a rendezvous in time (fig. 3.6). Independently of whether hosts are present to infect, the cercariae emerge from molluscs at the time of day that maximizes the probability of encountering the right host.

Théron and colleagues (Jourdane and Théron 1987; Pagès and Théron 1990; Combes et al. 1994; Théron and Combes 1995) have shown that in nature, natural selection can adapt populations of the same schistosome species to two different host species, even if the latter have different activity patterns and different relative abundances in different sites. Thus, in regions where *Schistosoma mansoni* can infect both humans and black rats, there is selection for peaks of emergence frequency in the middle of the day in sites where humans are the principal host and peaks in the late afternoon where rats are the main host.

Emergence rhythms adapted to maximize encounters with hosts have been selected for in almost all trematode species. They are particularly pronounced in cases in which the hosts the cercariae are "trying" to penetrate also have very marked rhythms of water use, as do the terrestrial mammals

Fig. 3.6. Rhythms of emergence from molluscs by cercariae of three *Schistosoma* species (the distance between two successive lines represents 10 percent of the total daily emission). (After Théron 1984.)

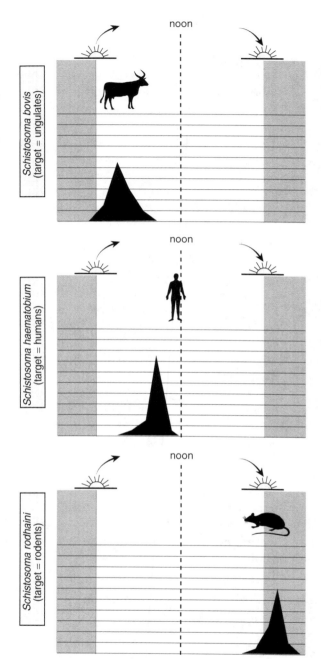

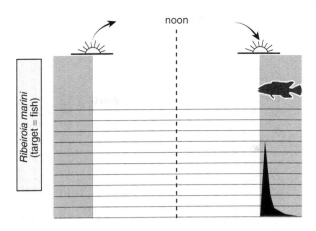

Fig. 3.7. Emergence   59
rhythm of cercariae of
the trematode *Ribeiroia
marini*, whose cercariae
penetrate freshwater
fishes.

(ungulates, humans, rodents) I have just discussed. However, there are also emergence rhythms in cercariae that penetrate aquatic hosts, such as fishes and insect larvae; in fact, these species have activity rhythms that accord with the fact that there is a higher probability of encountering cercariae at some times than at others. For example, consider the trematode *Ribeiroia marini,* whose adults parasitize fish-eating birds and whose metacercariae are produced in fishes. The cercariae emerge from molluscs just after sunset (as do those of *S. rodhaini*); this timing probably coincides with the period when the fishes, after a diurnal swimming period, rest immobile on the bottom, close to molluscs (fig. 3.7).

Arrhythmic emergences are rare, but tellingly they are found in species that encyst on plants, which of course do not have activity cycles. Thus, the cercariae of the sheep liver fluke, *Fasciola hepatica,* emerge irregularly at any hour of the day or night (fig. 3.8). The amphibious snail intermediate host, in which the cercariae are produced, emits them in water. They swim a short distance and usually encyst as metacercariae on plants, where they live for months and are ingested by sheep when the water level falls.

Rhythms selected to open the encounter filter exist in parasites other than trematodes. Best known are those of microfilariae, that is, the larvae of filarial nematodes found in blood of humans or other parasitized animals. In general, microfilariae, which must be absorbed by a biting insect to survive, "climb" to the most peripheral blood vessels, just under the skin, at the hour when the target insect is most likely to take its blood meal.

Adult males and females of the nematode causing elephantiasis (*Wuchereria bancrofti,* cited earlier) do not live in blood vessels but in the

Fig. 3.8. Arrhythmic emergence pattern of cercariae of the trematode *Fasciola hepatica*. (Data from Bouix-Busson et al. 1985.)

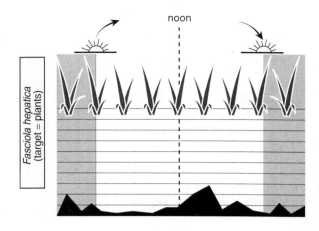

lymphatic system. In major cases, obstruction of the lymphatic vessels and eventual bacterial infections lead to elephantiasis, a pronounced edema that generally affects the lower limbs and, in males, the scrotum. After copulation, the female *Wuchereria* continually liberates tiny larvae, the microfilariae, that have no adaptation allowing them to exit the host. These pass from the lymph into the bloodstream. Microfilariae have a lifespan of several weeks and are very numerous. If a mosquito of the genera *Aedes* or *Culex* ingests microfilariae during a blood meal, these escape being digested and actively pass through the digestive tract wall. Thus they reach the insect hemocoel (the cavity containing the blood), and about two weeks later they reach the mouthparts, from which they are injected into a new host individual.

What is remarkable in these *Wuchereria* nematodes (and others with a similar transmission mode) is the way these microfilariae, although not participating actively in passage through the host integument, facilitate their being taken by mosquitoes. Natural selection has produced a double rendezvous strategy, in space and in time.

The need for a spatial rendezvous is understood if one compares the length of the biting apparatus of the mosquito to the volume of a mammalian body. It is evident that the mosquito can take blood meals only in the most peripheral blood vessels. Any microfilariae in deep vessels have no chance of being ingested by mosquitoes.

The need for a temporal rendezvous is tied to the fact that mosquitoes bite at well-delimited times. For *W. bancrofti,* the vector species are nocturnal, never bite before six o'clock in the evening, and cease all activity at dawn. Therefore microfilariae can be ingested only during nighttime.

The double rendezvous is precisely inscribed in the microfilarial genes. During daytime, microfilariae are nearly absent from cutaneous blood vessels. Conversely, degree of peripheral microfilaremia (indexed by the number of microfilariae per unit volume of blood) rises sharply at nightfall. The microfilariae gather every night in the cutaneous blood vessels, waiting, so to speak, for their vector to carry them over the frontier between different host species.

The Parasite Creates Information

Some freshwater mussels are parasites. In fact, it is only their larvae, glochidia, that in order to survive must attach themselves to fish gills, where they grow. Several weeks later, the young mussels detach, fall to the bottom of the river, and commence a free-living existence. In some mussel species, the glochidia are simply dropped into the mud by their mother and must "wait" for a fish to ingest them accidentally while feeding. Then they attach to the gill filaments. In other species, such as *Lampsilis,* the encounter with the fish is mediated by mimicry, but the exact means by which the mollusc detects the fish is uncertain (Haag, Warren, and Shillingford 1999).

At the moment of reproduction, the female mussels open their valves and the fringe of their mantle emerges. This fringe waves in the water, constituting a signal that draws the attention of fishes that are always on the lookout for prey. In *Lampsilis radiata,* this piece of the mantle is bifurcated, which makes it resemble a fishtail. In *L. ventricosa,* the extruded mantle resembles an entire small fish (fig. 3.9). Cheng (1970, 239), from

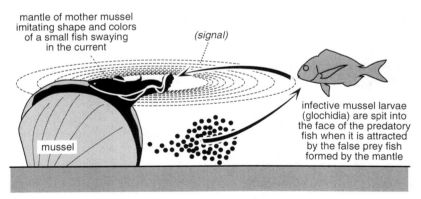

Fig. 3.9. The lure ("false fish") of freshwater mussels attracts a predatory fish, which is rewarded with a cloud of infective larvae spit into its face. The mussel larvae attach to fish gills and are temporarily parasitic there.

62 whose research I draw this description, wrote that the edge of the mantle is "amazingly fishlike; it bears an anterior head with an eye-spot and a posterior tail . . . it is partially bright-colored." When a predatory fish approaches this decoy, the mollusc spits out a cloud of thousands of glochidia, which had remained until then sheltered between the two maternal valves. These are aimed at the head of the fish, which becomes heavily infected while trying to capture the fish decoy.

In several trematode species there is a similar transmission episode. In *Leucochloridium paradoxum*, for example, there are not free-swimming cercariae, unlike in most other trematodes. The cercariae do not leave the parasite sporocyst in the host mollusc (a snail in the genus *Succinea* that inhabits freshwater shores) in which they were produced. Rather, they accumulate in one part of the sporocyst. They encyst there as infective metacercariae. The part of the sporocyst that contains them insinuates itself into one of the snail's tentacles, which acquires the shape, color, and movements of an insect caterpillar, so that this "sack of parasites" is swallowed by blackbirds, which become the hosts of the adult trematode. The blackbirds become infected with 200 to 300 parasites all at once (Bakke 1980). Darwin would surely have loved this spectacular illustration of natural selection.

These examples of the mussel whose mantle resembles a prey (fish) and attracts a predatory fish, and of the trematode that makes a snail antenna mimic a caterpillar, are exceptional. In fact, few parasites themselves create information that attracts their hosts. I could cite several other examples, notably that of cercariae that are eaten because they resemble mosquito larvae, but there are not many such cases. More often, as I will show, the information is created by the host that carries the infective stage, by a process of extending the phenotype.

The Parasite Forces Its Host to Create Information

If a persistent intimate interaction links two species with different genomes, they nevertheless constitute a union of two living objects functioning according to certain basic rules. From the smallest virus to the elephant, the constitutive molecules of living entities belong to certain biochemical families. Moreover, the genetic code, down to its tiniest details, is also the same for all species, proving that there exists a fundamental unity of life. These facts show that all life forms without exception are "cousins" of one another, even if their common ancestor existed very long ago.

Therefore, when individuals of two species, even distantly related, find themselves associated intimately for a long time, it is possible for the genes of one to send signals to the phenotype of the other. This is an illustration of what Richard Dawkins has called the *extended phenotype* (see chap. 1), signifying by this term that the expression of the genome of one of the partners extends beyond its own phenotype to the phenotype constructed by its partner's genome.

In theory, this meddling can occur in both directions, with the host genome affecting the parasite phenotype or the parasite genome affecting the host phenotype. In practice, it is the latter sort of extended phenotype that has usually been described—the parasite genome affects the host phenotype. These are usually called *genetic manipulations,* and they are a matter of parasite adaptations that give them additional weapons in the two arms races.

In the "encounter" arms race, the parasite can manipulate the host morphology or behavior, resulting in increased likelihood of transmission to the next host in the life cycle. I have shown that the parasite is not very sensitive to information coming from its host, but the parasite creates signals (that is, information) by the intermediary of its host. In the "compatibility" arms race, the parasite can modulate host immunity by secreting immunosuppressive molecules, and it can also provoke host ontogenetic changes that modify the parasite's habitat.

In many life cycles, the parasite "jumps" from one host to the next by being eaten along with the former host. It is of course in the interest of a potential prey item not to be consumed, but this would be inimical to the parasite, as the infective stage would then be stuck in this prey species. If the predator takes its meals haphazardly among the population of potential prey individuals, the parasite will be transmitted with a certain probability p, depending essentially on the distribution of parasites among the prey population and the frequency of predation events. However, if the predator tends to capture preferentially prey individuals that carry the parasite, the probability will exceed p, even if the number of parasites and the number of predation events remain unchanged.

This is obviously a great situation for natural selection. If there exist in the parasite population individuals able to cause p to increase, their genes will be selected for. The genes in question are "manipulative" genes, genes that affect the extended phenotype. Researchers have increasingly detected, if not these genes themselves, then at least their effects, and we know they act in various ways, according to the particular life cycle in question. I cite several examples from among trematodes and tapeworms.

The oystercatcher *Haematopus ostralagus* is a beautiful bird with rose-colored feet and a long, coral-colored beak, the top of the body black and

64   the underside white. It feeds on shellfish and crustaceans found on boulders and in mudflats. It is especially common on the shores of boreal seas, but it is not uncommon in the Gulf of Lion, in the Mediterranean. There, in the lagoons of the Camargue, Pierre Bartoli discovered surprising details of the transmission of an oystercatcher parasite, the trematode *Meiogymnophallus fossarum*, which I have already mentioned while discussing the spatial specialization of cercariae. The following description is based entirely on his research (see Bartoli 1973a, 1973b, 1978).

The life cycle of *M. fossarum* begins with larval multiplication in the bivalve *Scrobicularia plana* (and only this species). The cercariae arising from this multiplication are liberated in the water in prodigious numbers: Bartoli estimates that the density of *S. plana* carrying the parasite can reach 10 per square meter over dozens of hectares and that more than five million cercariae can be produced in each bivalve.

Once released from this "productive" bivalve, the cercariae install themselves in a second bivalve. Here the degree of specialization is less pronounced than in selection of the first host. Several species may be used. Two, however, play a dominant role—the cockle *Cerastoderma glaucum* and the clam *Venerupis aurea*. To understand what happens next, recall some peculiarities of bivalve anatomy (fig. 3.10). The two valves of the

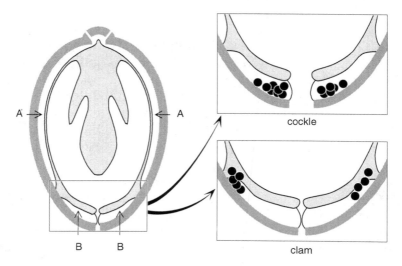

Fig. 3.10. (*Left*) Cross-section diagram showing bivalve anatomy. (*Right*) Location of *Meiogymnophallus fossarum* metacercariae in cockles and clams. The silhouettes of the two bivalves are represented identically to simplify comparison. (After various articles by Pierre Bartoli.)

cockle are connected by a dorsal hinge. The shell is secreted by the peripheral edge of the mantle. Between the mantle and the shell is a space known
as the extrapallial space, which is divided into two parts by the line where
the mantle is attached to the shell. The part next to the hinge is the general
extrapallial space (fig. 3.10, A); the part toward the exterior is the peripheral extrapallial space (fig. 3.10, B).

In either the cockle or the clam, the cercariae of *M. fossarum* penetrate
various body parts, especially the labial palps and gills. Subsequently, they
emigrate through the tissues and end by falling into the peripheral extrapallial space, where they become free metacercariae (not encysted, as is usual
for trematodes). This is the region where the bivalve shell is actively growing. Any metacercaria that does not reach this extrapallial space perishes.

From here, an important difference appears, depending on which host
species the metacercaria is found in. In the cockle, the metacercariae mass
in the most marginal region of the peripheral extrapallial space.[17] In the
clam, they are also found in the peripheral extrapallial space, but they
occupy the most dorsal part. Furthermore, in the cockle the metacercariae
remain in the extrapallial space, whereas in the clam vesicles form that
gradually encase the metacercariae toward the interior of the thickness of
the mantle. The most astonishing facts are yet to come; they concern the
consequences of this invasion for the subsequent transmission of the parasite to the bird.[18]

In the cockle, the presence of great numbers of metacercariae exactly at
the place where the shell forms has repercussions on the functioning of the
tissues that build the shell. These tissues, although they have no visible
defense mechanism against the parasite, produce villosities that are not
found in healthy animals. The shell, which arises from the precipitation of
calcium salts contained in the liquid in the peripheral extrapallial space, is
not produced normally in parasitized cockles. Gaps appear where usually
the two valves would join cleanly. In a healthy, closed animal, the body is
entirely hidden by the shell. In a parasitized animal, by contrast, the irregularities of the edges of the shell allow openings to the interior.

In the clam, these irregularities in the closure zone of the two shells are
not produced, which is unsurprising because the parasite is located outside
the zone where calcium salts are precipitated. However, another anomaly is
seen (fig. 3.11). Bartoli first noticed that the metacercariae were unequally
distributed around the perimeter of the peripheral extrapallial space. The
majority occupy the zone surrounding the mollusc siphons, especially in
the region just behind the siphon. This has the effect of pushing the
siphons toward the ventral region, following a rotation of about 45°. And

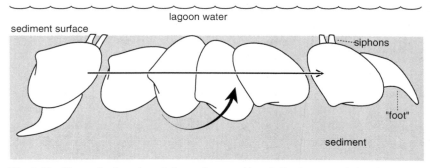

lagoon water

sediment surface

siphons

"foot"

sediment

Fig. 3.11. Changed orientation of clams after parasitism, facilitating transmission. (After Bartoli 1978, modified.)

finally, this change causes the animal itself to turn so that the siphons continue to open at the interface of the sediment and the water.

These modifications, simultaneously morphological and behavioral, that affect parasitized cockles and clams, although different in the two bivalve species, have similar consequences with respect to consumption of the bivalves by the oystercatcher. Among the cockles, the animals with the poorly sealed shells are easier to open. Among the clams, the turning of the animal exposes the gaping pallial cavity toward the surface of the sediment. Bartoli (1978, 27) writes, "In this new posture, *Venerupis aurea* becomes an extremely easy prey for *Haematopus ostralagus*."

The evolution of such traits, which increase the probability that a parasite (whichever species it is) will be transmitted from one host to another, is now called *favorization* (Combes 1980, 1991a). What mechanisms are in play? In other words, how, in the preceding example, does *M. fossarum* succeed in manipulating the morphology and behavior of its hosts? It is important to know that, in the majority of the many life cycles in which the existence of favorization has been demonstrated or suspected, knowledge of the mechanisms is still very limited.

For *M. fossarum*, we are initially tempted to believe that the effect of the parasite is only mechanical because the effect seems proportional to the intensity of the infection, which is often enormous. However, this is far from constituting a proof. In the cockle, for example, the metacercariae are not generally in contact with the host tissues. It is therefore possible that the parasite directs toward the tissues that produce the shell a molecule that modifies the composition of the extrapallial liquid, or even the functioning of the cells controlling the deposition of calcium compounds. Whatever the mechanism, favorization in *M. fossarum* is one of the best examples known in which the phenotype of the parasite extends into the phenotype of its host. To my knowledge, this example is moreover unique among symbioses in the sense that the same parasite is able to manipulate two host species in different ways but to achieve the same result.[19]

From Blind Fish to Phantom Flowers

The life cycle of *Diplostomum* incorporates three hosts: a gastropod mollusc, a freshwater fish, and a fish-eating bird. The transmission from the fish to the bird is achieved when the latter eats the former. Here it is a matter of very common trematodes that sometimes cause sudden mortality when trout or carp are farmed.

The swimming cercariae emerge from molluscs, then penetrate the skin of fishes, in which they become metacercariae that are mobile (not encysted) and measure 2 to 3 millimeters long. Wherever the cercaria enters, the metacercaria wends its way through the tissues of the fish in the direction of the head and finds the eye in less than 24 hours. Depending on the parasite species, the metacercariae localize in the crystalline lens or other regions of the eye (aqueous humor, vitreous humor, retina). For instance, metacercariae of *Diplostomum spathaceum*, the best-known species in this genus, are always found in the crystalline lens, where they cause an opacity comparable in every way to a cataract. Metacercariae of the closely related species *D. gasterostei* are found in the retina. As soon as the number of parasites per eye surpasses four or five, the vision of the fish deteriorates, as is proven by its difficulty capturing prey. If the number of parasites increases still further, not only is the fish blinded but it also tends to swim closer to the surface than healthy fish do. Although the infection entrains an immune reaction on the part of the fish, the rapid localization of the metacercariae in the eye protects them, as this organ is but slightly immunogenic.

So far there have not been field predation tests affirming that birds more easily capture fish infected by *Diplostomum* than they do other fish. However, the fact that the parasitized fish are blind and swim at shallow depths (perhaps because they are seeking well-lit areas) allows the strong inference that the localization of the metacercariae in the eye can be interpreted as a phenomenon favoring transmission. If this interpretation is correct, it means that information causing migration in the direction of the host eye has been selected for in the *Diplostomum* genome.

Favorization is known in many other trematode life cycles. That of *Microphallus papillorobustus* includes, in succession, a mollusc (*Hydrobia*), an amphipod of the genus *Gammarus*, and a bird. The bird (for example, a gull) becomes infected by eating the amphipods, which are small crus-

68   taceans of fresh or brackish water. The metacercariae concentrate in the amphipod nerve centers, and striking behavioral changes are associated with this fact (Helluy 1981, 1982). Parasitized amphipods tend to be attracted by light, to prefer the surface of the water rather than its depths, and above all to move vigorously when threatened; these traits are all exactly the opposite of those manifested by normal individuals. Predation tests have shown that parasitized amphipods have at least four times the probability of being eaten by a gull than a healthy individual has, for the simple reason that it is easier for the bird to detect active prey on the surface than immobile prey in deeper water.

The tapeworm *Anomotaenia brevis* has a life cycle with two hosts, an ant inhabiting dead wood and belonging to the genus *Leptothorax*, in which the infective larval stage (called a cysticercoid) is found, and a bird (mainly the great-spotted woodpecker) that houses the adult. The ants are infected as larvae when worker ants bring them, as food, the proglottids of the tapeworm, discovered while foraging. Parasitized ants change in many ways: the workers are unusually small, their heads and feet are short; individuals originally destined to be queens instead develop into a stage called *intercaste*, intermediate between queens and workers (Plateaux 1972; Péru 1982). But most strikingly, parasitized individuals are a golden yellow, in contrast to the brown of normal individuals. Further, when an animal approaches the anthill, these parasitized individuals stay on the nest instead of fleeing. There is therefore a double modification of appearance and behavior that favors ingestion of the infected ants by the woodpecker, the final host. The differences between healthy and parasitized individuals are so pronounced that it is possible that entomologists occasionally have described them as belonging to different ant species. Janice Moore (1995), discussing this sort of manipulation, metaphorically subtitled an article "When an Ant Is Not an Ant." I should add an interesting detail: the cysticercoids are attached to the wall of the ant digestive tract, behind the gizzard, and the presence of a single individual suffices to generate the modifications. It is therefore doubtful that the loss of energy incurred by the ant is solely responsible for the production of the abnormal phenotype, as a result of the changes the energy loss causes in the immature individual. Here as in many other instances, the parasites may emit molecular signals that cause the changes in the host.

The life cycle of *Dicrocoelium* follows a disconcerting path. As in all trematodes, the first host is a mollusc, but, whereas other trematodes usually use an aquatic mollusc, *Dicrocoelium* parasitizes *Helicella*, a small terrestrial snail. The second host is an ant, and the third is a sheep. The first passage, from snail to ant, poses problems at the outset because the cercariae produced inside the mollusc cannot use water to swim to the next host. This problem is solved by the intermediary of small mucus globs emitted by the mollusc; the cercariae are found in these globs. These droplets attract ants, which ingest them and thus become carriers of metacercariae.

At this point the second incongruity of this life cycle arises: sheep do not eat ants, or at best ingest them accidentally while browsing. At this step, the probability of the life cycle being completed is therefore very slim, unless parasitized ants were to climb up grass blades, in which case the probability of reaching a sheep gut would be greatly increased. A manipulation comparable to that in the *Microphallus* of amphipods has arisen. One of the metacercariae lodges in the sub-esophageal nerve ganglia that control the movement of the ant mandibles and feet, with spectacular consequences: for starters, the parasitized ant climbs to the end of the grass blade. Then it grabs the blade with its mandibles, with its head therefore aimed down. It is not an overstatement to say that all this happens as if the ant was "awaiting" the sheep. However, during the hot part of the day, to remain grasping the top of a grass blade would be suicidal because of the heat. If the ant is not eaten, it climbs down to the ground again and does not climb up on the plant again to wait for the sheep until the following night. In comparison to the amphipod parasite, the difference in this life cycle is slight. It consists mainly of the fact that, in the *Dicrocoelium* cycle, a single metacercaria lodges in the neural ganglia of the ant, while many metacercariae do so in the *Microphallus* cycle.

The *Dicrocoelium* life cycle is an excellent entrée into the concept of kin selection (Sigmund and Hauert 2002). This case seems to indicate that the metacercaria that lodges in the sub-esophageal ganglion of an ant pays for this behavior with its life, because it apparently dies, so to speak, in the course of duty (it does not become infective). It is thanks to this metacercaria that the sheep is infected, but this individual metacercaria will be the only one not to transmit its genes. Paradoxical? No, because the preceding reasoning is not quite right. In fact, all the cercariae produced in an individual mollusc arise (except in the rare case of multiple infection) from the multiplication of a single miracidium, so they all have exactly the same genes. Therefore, the sacrifice of a metacercaria by virtue of its lodging in the ganglion is not really a sacrifice in the genetic sense, because its genes are transmitted by its identical sisters. This is a case of kin selection, and it is an excellent example of how an individual can be of little consequence to evolution; it is the survival of genetic information that is selected for.

70 Another less-known trematode life cycle, studied by W. Carney, also involves ants, with features and consequences quite different from those in the *Dicrocoelium* case. This is the cycle of *Brachylecithum mosquensis,* whose cycle passes in order through a terrestrial snail, a carpenter ant of the genus *Camponotus,* and birds that become infected when they eat the ants. The ants infect themselves by consuming mucus droplets containing cercariae, as in *Dicrocoelium.* Luc Passera (1975, 239) describes what happens next:

> As in the small fluke, a metacercaria encysts in the cephalic nerve centers, but in this species it chooses not the sub-esophageal ganglion but the region that innervates the ocelli, eyes, and antennae. This induces a behavior that favors the capture of *Camponotus* by birds, but one that differs greatly from the mandibular spasm of *Formica* (the ant in the *Dicrocoelium* case). Here, the parasitized workers leave the undergrowth where the healthy workers are stationed and slowly walk in circles or remain immobilized on rocks along watercourses, in broad daylight. Furthermore, these workers are seen to be obese; the abdomen is distended and the intersegmental membranes become an opaque white. This coloration and the very slow movement of the ants in sunlight far from any vegetation renders them visible at a glance.

It seems that, among other effects, the ant is no longer able to assess light intensity properly.

In the life cycles of many tapeworms, sheep are the intermediate hosts and must be eaten by a carnivore for the parasite to reach adulthood. For example, *Multiceps multiceps* is found as an infective cysticercus in sheep and as an adult in dogs and wolves. Two adaptations favor the capture of infected sheep. On the one hand, the cysticercus is of the "budding" type; that is, it grows substantially while multiplying the number of parasites by asexual reproduction. On the other hand, the cysticercus is found in the brain, in the cerebellum, where its presence induces a sort of "whirling disease." It is obvious that, if a wolf pack chases a flock of sheep, the pack will capture individuals moving in circles, thereby transmitting the tapeworm.

Ligula intestinalis is a pseudophyllidean tapeworm whose life cycle has three successive obligatory hosts (fig. 3.12). First is a copepod that becomes infected by ingesting the free-living stage, the coracidium (which hatches from an egg); the coracidium then transforms to a procercoid. A freshwater fish then eats the copepod and procercoid; in the fish, the latter becomes a plerocercoid. Finally, a bird becomes infected by devouring the fish; in the bird, the plerocercoid becomes an adult tapeworm, inhabiting the intestine.

The three successive hosts are exploited very unequally. Although the copepod and bird barely appear to suffer from a *Ligula* infection, the fish is

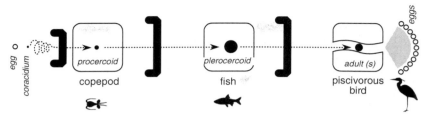

Fig. 3.12. Life cycle of the pseudophyllid tapeworm *Ligula intestinalis*. Observe that three predation acts intervene in succession during this cycle.

gravely afflicted; it is the host that contributes by far the most to the parasite energy budget. During its spell in the body cavity of the fish, the plerocercoid grows enormously, reaching a third of the body weight of the host. This monopolization of metabolites is at the expense of the reproduction of the parasitized fish, whose gonads remain undeveloped, whether male or female. Whatever the mechanism at play (simply rerouting of metabolites or, more likely, a molecular secretion that modifies the fish endocrine function), this is a characteristic example of an extended phenotype.

The parasite gains two advantages: on the one hand the acquisition of resources that allow strong growth in anticipation of the development to occur in the third host, and on the other, a weakening of the fish that enhances the probability a bird will see and capture it. Closely related parasite species can have quite different adaptations. In another pseudophyllidean tapeworm, *Schistocephalus solidus,* whose life cycle formally resembles that of *L. intestinalis,* the fish host gonads are not dysfunctional, suggesting that the mechanisms of manipulation generated by natural selection differ from those of *Ligula.* It is notable that, within *Ligula* itself, different populations can have different manipulative effects (Olson et al. 2002).

In some cases, manipulation of host behavior is linked to a disruption of the molecular neurotransmitters that govern nervous system activity. In rodents, stress provoked by the sight of a predator entrains the liberation of analgesic substances that prevent panic reactions (stress analgesia) and therefore provide the best possible chance that the potential prey does not become a victim. These substances are part of the defense mechanism and are liberated at synapses in the brain. Kavaliers and Colwell (1994, 1995) have studied the influence of various parasites on stress analgesia. The latter can be lost under the influence of parasites, and thus predators more easily capture parasitized rodents.

Plant phenotypes can be manipulated by animals. I cited the example of gall makers in chapter 1. Plant phenotypes can also be manipulated by the genes of fungal parasites, which can even make phantom flowers form to assure their transmission.

The fungus *Monilinia vaccinii* grows on blueberries and causes the berries to dry. These "mummified" berries fall to the ground, and the fungus undergoes sexual reproduction there to produce spores. These spores, carried by wind in springtime, infect young blueberry leaves. On the leaves, the fungus undergoes asexual multiplication to produce another generation of spores (conidia) that are carried to the blueberry flowers. The fungus achieves this transport by trickery, by making the infected leaves attractive to the normal insect pollinators of blueberry. The attraction is produced by modifying the leaf phenotype (Batra and Batra 1985). The leaves produce nectar and volatile substances (with the scent of fermented tea, according to the authors) that attract the insect, but the most astonishing fact is that the infected leaves reflect ultraviolet light, just as do "real" flowers (while the surrounding healthy leaves absorb ultraviolet light). In the flowers, the infection occurs when insects place conidia on the stigmas, just as they do pollen grains. This infection leads to mummification of the fruit. The molecular mediators implicated in the color change in these plants are still poorly known (Roy 1994).

The preceding examples are but a sample of the many research results available today. I have chosen them for their illustrative value but also because they allow us to see that very different parasites use similar means of manipulation. This is a new example of the phenomenon of evolutionary convergence, which may occur when different species must solve similar problems (transmission to a predator, in this case). These examples all lead us also to pose this question: Given that consumption of manipulated and infected hosts is disadvantageous to predators, why hasn't selection occurred in predators for new behaviors that cause them to pass up prey with modified behavior or appearance? Several possible answers must be considered. The simplest is that selection for these new behaviors would be even costlier, in terms of reproductive success, than being parasitized. For example, it is probably more advantageous for a wolf to conserve its energy by capturing a sheep stumbling in circles than to avoid contamination by cestodes. We can guess that this answer is valid so long as the parasites are not too virulent. Possibly this is one of the facts that "constrains" the evolution of parasite species, or at least some parasite species, to limit their virulence.

The Astounding Biology of Bacteria in the Genus *Wolbachia*

Until now I have emphasized manipulations that result in consumption by a predatory "definitive host" of a parasitized "intermediate host" prey.

I have noted that host manipulation can entail other sorts of processes. The *Wolbachia* bacteria provide an example.

Wolbachia are bacteria whose biology has been intensively studied during the last decade as their surprising manipulative properties have been revealed. They are found mainly in crustaceans and insects but also in filarial nematodes. They are located in the cytoplasm of host cells and cannot be cultured separately. *Wolbachia* do not produce a visible pathogenic effect on the morphology or anatomy of their hosts, but they can greatly influence host reproduction.

A crucial fact is that *Wolbachia* are transmitted vertically, from "parents" to "offspring," and only by host females. When the host female reproduces, *Wolbachia* enter the cytoplasm of the oocytes and, after fertilization, spread throughout the cells of the embryo. By contrast, if *Wolbachia* are found in a host male, there is not transmission by the spermatozoa, probably because sperm do not contain enough cytoplasm to harbor the bacteria. Therefore only *Wolbachia* found in host females transmit descendants.

Logical consequences of this transmission mode are as follows: (1) If *Wolbachia* were able to transform males into females—that is, to make them produce oocytes instead of spermatozoa—they could then be transmitted by these "feminized" males. (2) If *Wolbachia* were able to prevent healthy (uninfected) females from reproducing, they would favor the descendants (infected) of females that carry the bacteria, by reducing intraspecific competition suffered by the latter individuals. What is most remarkable is that *Wolbachia* can, in fact, accomplish both of these manipulations.

To bring about the transformation of male hosts into females, the *Wolbachia* use several strategies. The most widespread is that *Wolbachia* have feminizing genes that act on the sexual glands of the male hosts and transform them into ovaries![20]

To confer an advantage on offspring of infected females at the expense of offspring of uninfected females, *Wolbachia* induce the death of embryos when an infected male mates with an uninfected female. We say that *Wolbachia* has caused a *cytoplasmic incompatibility* between the sperm and the egg it has penetrated. *Wolbachia* thus contrive to prevent the birth of uninfected hosts.

There exist other striking traits in the biology of *Wolbachia*. For example:

1. *Wolbachia,* which often parasitize insect parasitoids whose larvae develop in the bodies of other insects, can pass from the victim to the parasitoid and vice versa, as has been shown by Vavre et al. (1999) from the molecular phylogeny of *Wolbachia* stocks.

2. *Wolbachia* can become indispensable mutualists in some hosts: Dedeine et al. (2001) have shown that, in a small wasp parasitoid, *Asobara tabida*, *Wolbachia* have become necessary for oocyte maturation in the females. In the absence of *Wolbachia*—for example, if they are experimentally eliminated by antibiotics—female *A. tabida* cannot reproduce.

3. In fruit flies, *Wolbachia* are even able to counteract the effects of a mutation that renders the insects sterile: Starr and Cline (2002) have proven experimentally that fruit fly females that carry this mutation become fertile again upon infection by *Wolbachia*.

In all of these cases, the bacterial genome exercises a strong control over the phenotype of the host by feminizing the males, or by killing embryos that it cannot infect, or by enhancing or even restoring the fertility of the females. Such passage of parasite genetic information to the host is one of the best-known illustrations of the concept of extended phenotype. Two of the most spectacular effects are depicted in figure 3.13. It is perhaps necessary to add, with F. Dedeine, that the astounding *Wolbachia* are evolutionarily closely related to the bacteria that gave rise to mitochondria of eukaryote cells, a fact that certainly spurs reflection.

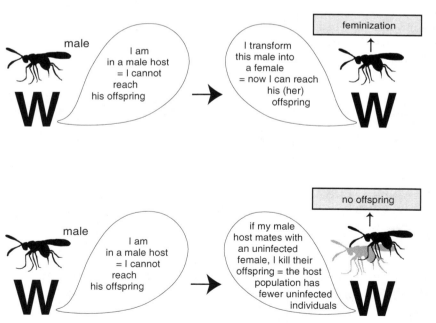

Fig. 3.13. Two manipulations by which *Wolbachia* modify the sex ratio of their hosts.

A Parasite Must Survive in Its Host

Escaping Host Defenses

Even if the parasite has won the first round (that is, it has detected or created enough information to encounter a host or has encountered a host simply by luck), it still has to establish itself in an environment (the host) that can defend itself. Hanley et al. (1996, 371) write that "host-finding and settlement by parasites can be analyzed using well-developed ecological and behavioral models for habitat choice and patch settlement. The difference, however, is that in parasite settlement, the habitat is able to fight back." The recognition of non-self and the triggering of highly elaborate defense mechanisms are vital processes for any living organism. One of the most important aspects of the parasite profession, therefore, is being able to survive and establish in a hostile environment. This is what I have called (chap. 1) selection for "survival genes." To this end, parasites use all sorts of means, such as hiding in organs that are not very immunogenic, or even inside protective cysts (seclusion); possessing surface molecules resembling those of the host (molecular mimicry); possessing receptor molecules able to adsorb molecules belonging to the host itself (molecular camouflage); or continually changing surface molecules (antigenic variation).

Among these possible mechanisms, those whose users succeed most frequently in avoiding the notice of the host defensive system have been selected by many parasite species. I give two examples here that seem at first blush to be very different but in which the strategies are in fact similar.

Beginning in the 1960s, we have known that host molecules such as blood-type antigens, antigens of the major histocompatibility complex, and even immunoglobulins (antibodies) can be adsorbed to the surface of certain internal parasites, rendering the latter nearly invisible to the immune system, which confounds them with "self." Some parasites directly express on their surface proteins or glucidic molecules that resemble the structure of host molecules, which makes them difficult to detect, as in the previous case. These strategies, which illustrate the phenomena of mimicry and molecular camouflage, respectively, are used by schistosomes of humans but also by many other parasite species of invertebrate as well as vertebrate hosts. The strategy of mimicking the host molecules and attaching these chemicals to their integument has been best demonstrated for glycolipids and glycoproteins of schistosomes. The presence of these mimetic molecules on the integument has been demonstrated experimentally in tests using two animals species, A and B (for example, hamsters and guinea pigs), both susceptible to the same schistosome species (for instance, *Schistosoma mansoni*).

76   First, host A is infected and host B is immunized against the blood serum of
A. Then schistosomes that developed in A are microsurgically transferred to
the circulatory system of B. These schistosomes are immediately destroyed,
even though B is typically susceptible to them. We conclude that the schistosomes that developed in species A were attacked because they were covered with molecules modeled on those of A.

Let us now compare the strategy of certain social parasitic insects to
those just outlined.

The wasp *Polistes atrimandibularis* parasitizes a near relative, *Polistes
biglumis* (here I am summarizing the work of Jean-Luc Clément and his
team, Bagnères et al. 1996).[21] Unlike the majority of social hymenopterans, *P. atrimandibularis* has no worker caste (and is therefore an obligatory social parasite). In the beginning of the summer, when the host
species has just founded new colonies, a queen of *P. atrimandibularis* that
was fertilized the previous summer enters the nest of a *P. biglumis* founder
queen. At the moment when this violation of the nest occurs, the host
queen is very aggressive (but not very effective in defending her nest),
while the parasite queen is hardly aggressive toward its host. In several
hours, everything changes: the parasite queen becomes increasingly dominant and begins to lay her eggs, while the host queen becomes submis-
sive. The usurper lays only eggs that produce sexual individuals (males
and females) and lays them in greater number than does the dethroned
queen, although the latter remains in the nest (she is useful because she
produces workers who rear the sexual offspring of the invader). Because
P. atrimandibularis produces only sexual offspring and no workers, it
"must" be parasitic and have its sexual offspring tended by workers of the
host wasp.

In social insects in general, discrimination between individuals of different species or even between individuals of the same species but different
colonies rests on olfactory or gustatory cues. Each species, or sometimes
each colony, has (to use the expression of Jean-Luc Clément and his col-
laborators) a chemical signature that allows discrimination between related
individuals and strangers. The species-specific or colony-specific signature
is produced by hydrocarbon molecules present on the cuticle in bees,
wasps, ants, and termites.[22] Much research has shown that associations
between individuals of two species are possible only if there exist camouflage mechanisms involving these cuticular hydrocarbons (see, for example, Bonavita-Cougourdan et al. 1997).

In *Polistes*, Bagnères et al. (1996) compared the chemical signatures of the
host and the parasite at different stages in the invasion. Before the invasion,

the chemical signatures are totally different, as is shown by chromatographic analysis of the cuticles (fig. 3.14, *left*); the main difference resides in the nature of the hydrocarbons, which are unsaturated molecules in *P. atrimandibularis* and saturated molecules in *P. biglumis*. Just after the invasion, in late June, the unsaturated hydrocarbons of *P. atrimandibularis* disappear and are replaced by molecules resembling those of the host, thus making their chemical signatures more similar (fig. 3.14, *right*). Finally, when the sexual offspring and workers of the host queen are born in late July, the signatures of the parasite queen and the host queen are completely indistinguishable.

From this point, the outcome is easy to guess: the descendants of the *P. biglumis* queen cannot distinguish between their mother and the parasite queen and place themselves in the service of the sexual offspring of the parasite, which assures her reproductive success. In the end, the descendants of the parasite queen are more numerous than those of the host queen. According to Bagnères et al. (1996), the adaptive changes in the composition of the integument hydrocarbons of the parasitic wasp are

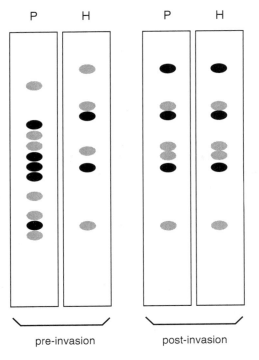

P   H   P   H

pre-invasion   post-invasion

Fig. 3.14. In several hours, the "chemical signature" of the parasitic wasp (P) mimics that of the parasitized wasp (H). The profiles are different before the invasion but practically identical after it, which allows the parasite not to be recognized as "non-self" by the host. The vertical rectangles represent chromatographic results. The blots represent hydrocarbons of different molecular weights; the dark dots represent the most abundant molecules. (After Bagnères et al. 1996, very simplified.)

78   explained either by activation/suppression of genes coding for enzymes
       implicated in hydrocarbon synthesis or by the adsorption of hydrocarbons
       borrowed from the host. It is evident that there is no fundamental differ-
       ence between the strategy of molecular mimicry of *Schistosoma* (a flat-
       worm) and that of *Polistes* (an insect).

       Another very effective process (although a dangerous one; we will see
       why in a moment) consists of producing molecules that intervene di-
       rectly in the immune mechanisms of the host, depressing or suppressing
       them. Such immunosuppressive mechanisms are known in the majority
       of parasites. For example, schistosomes fabricate a substance (SDIF, for
       schistosome-derived inhibitory factor) that limits the proliferation of
       host lymphocytes, the liberation of defensive products by the mastocytes,
       and the activation of the platelets in humans. Schistosome proteases
       cleave antibodies that attach to the surface of the parasite when invasion
       occurs, which leads to the liberation of a small peptide molecule that is
       active against macrophages. Research on *Schistosoma mansoni*, the agent
       of intestinal schistosomiasis, has shown the presence of peptides such
       as beta-endorphin, which are produced in humans by the anterior region
       of the pituitary gland. These molecules interfere in a complex manner
       with the immune system mechanisms by immunosuppression, especially
       by reducing antibody production. In other parasites, such as the agents
       of Chagas' disease in South America (*Trypanosoma cruzi*) and of leish-
       maniasis (*Leishmania major*), the cascade of conversions and polymer-
       izations that characterizes the activity of complement (which plays an
       essential role in immunity) is stymied by specific molecules emitted by
       the parasites.

       Why is this sort of process dangerous? Simply because to depress host
       immunity is often, for a parasite, to open the door to other parasites. Among
       invertebrates, I can cite the example of some trematode species that do not
       develop in a particular species of mollusc unless a different trematode species
       has paved the way by modifying host defenses. For instance, *Schistosoma bovis*
       does not use the gastropod *Bulinus tropicus* as intermediate host unless the
       individual *Bulinus* has already been infected by another trematode of the
       genus *Calicophoron* (Southgate et al. 1989). In vertebrates, we need only recall
       the dramatic example of the human immunodeficiency virus (HIV), which
       induces a sharp lowering of natural defenses and thus entrains the prolifera-
       tion of "opportunistic" parasites that, without the virus, have only the slim-
       mest chance of infecting and becoming pathogenic (for example, the fungus
       Candida albicans, protozoans of the genera *Cryptosporidium*, *Toxoplasma*, and
       Pneumocystis, microsporidia, bacteria, and viruses).

Similar to manipulation, although the strategy is slightly different, is use by the parasite for its own benefit of host molecules that are supposed to combat it. I give as an example interleukin-7, a molecule manufactured by the immune system to fight parasites and that, paradoxically, enhances the infection of mice and probably humans by the trematode *Schistosoma mansoni*. In fact, in mice deficient in producing interleukin-7, these schistosomes remain dwarfed and do not mature sexually (Wolowczuk, Roye et al. 1999; Wolowczuk, Nutten et al. 1999).

Reconstructing Its Habitat
Niche Construction

The parasite, in certain cases, is not content simply to neutralize host defenses. It can also manipulate its living habitat toward its own ends as much as possible.

The notion of "constructing a niche" has been proposed by several researchers (Laland, Odling-Smee, and Feldman 1996, 2000; Odling-Smee, Laland, and Feldman 2003). The idea is that all species modify the environment they live in and therefore the selective pressures imposed on them by this environment. For example, earthworms influence the structure of the soil and must be adapted to the environment as it is modified by their mechanical activity and their secretions.[23]

Niche construction by parasites does not have all the evolutionary consequences that it has in free-living species. In free-living species, in effect, the modified niche can be bequeathed to descendants (young earthworms are born in an environment modified by their parents). For parasites, the fact that the environment (host) is mortal means that the phenotypic changes in the host induced by the parasite cannot be passed on to its descendants except for a very short while. Even such a short-term bequest need not be advantageous unless the descendants parasitize the same host individual, which is generally not the case. Despite this problem, parasites are nonetheless superb constructors (or, more precisely, reconstructors) of their niches.

The Trichina Case

The life cycle of the trichina (the nematode *Trichinella spiralis*) is intriguing in the sense that this parasite never sees the light of day. There is no free-living stage in its life cycle because the trichina passes endlessly from one mammal to another so long as these eat one another. For example, the parasite can pass from a rat to a pig if the pig eats the rat, then from the pig to the rat if the rat eats pork scraps. It can also pass from pig to a

80   human if the human eats undercooked pork. Obviously, humans are a parasitic dead end in this chain.

When an animal eats a contaminated meal, it ingests the parasite in the form of tiny larvae that rapidly become adult males and females in the digestive tract. After copulation, the female is ovoviviparous, liberating many larvae, which traverse the intestinal wall and lodge in muscles.

The most interesting aspect of the life cycle, however, transpires in the muscle. Surprisingly, the trichina larva is not content simply to inhabit muscle tissue; it penetrates a single muscle cell. It is thus a metazoan parasite of a cell (like *Polypodium* in chap. 2). And then the manipulation begins, because in just three weeks the trichina larva completely reorganizes its habitat (see Despommier 1993). By molecular signals that are still far from completely elucidated, the trichina larva induces in its host cell (1) disappearance of actin and myosin filaments; (2) growth in size of nuclei (muscle cells are multinuclear); and above all, (3) differentiation around the cell of a mass of capillaries, known as a circulatory rete, comparable in every way to the placenta that forms at the junction of the mammalian embryo with the maternal uterus. This angiogenesis (formation of blood vessels) proves that the genes of the minuscule trichina larva govern developmental processes that are normally exclusively controlled by mammal genes. It is likely (as suggested by the growth in size of the nuclei) that the trichina genes selectively activate certain genes of the host cell and that it is these genes that induce transformation of surrounding tissues.

Galls and Their Makers

The galls of plants, already mentioned in several places in this volume, are another classic example of an environment modified by the extended phenotype of parasites. Most often, the galls form around insect larvae—flies (Cecidomyidae) and wasps (Cynipidae). The extended phenotype crosses the frontier separating animal and plant kingdoms because the insect manipulates plant tissues. The mechanisms by which an insect induces a modification of plant tissues are still poorly known. However, the cause-and-effect relationship between the parasite and the host bearing a gall has long been recognized. Malpighi in the seventeenth century was the first to detect this relationship.

Taxonomically close insect species can induce formation of galls that are morphologically very different. More remarkably, one species of gall maker can induce the formation of several different kinds of galls on the same host. Even the same genotype can cause several distinct manipulations of the same phenotype.

Westphall, Bronner, and Michler (1987) cite two cases, one of which is associated with sex, the other with the mode of reproduction. With respect to sex, females of the wasp *Mikiola fagi* cause large red galls in their hosts, whereas males cause small green ones. As for mode of reproduction, some insects have two successive generations per year, one parthenogenetic (with females producing offspring without males) and the other sexual (with males and females mating before the females reproduce). The cynipid wasp *Neuroterus quercusbaccarum* produces small spherical galls on oak leaves when reproducing parthenogenetically and flattened, elongated galls when reproducing sexually.

We also observe that certain galls, induced by fungi, can attain a diameter of three meters (for *Taphrynia insitisiae* on plum trees) and that formation of other galls entails a third partner. Galls caused by the wasp *Lasioptera rubi* and several other species are internally carpeted by a fungus introduced by the female wasp when she oviposits; the fungus aids larval nutrition.

Similarly, to reorganize their habitat, certain nematodes that live in plant roots induce the differentiation of giant feeding cells or even cause cell walls to dissolve (Davis, Hussey, and Baum 2004).

The Case of Sacculina carcini, *a Parasitic Barnacle*

Parasitism of crabs by *Sacculina carcini,* a parasitic barnacle, is the host–parasite association found in every parasitology textbook. This association between two crustaceans, known since the beginning of the nineteenth century, is usually used to demonstrate to what degree a parasitic lifestyle can lead a species to differ from its ancestors. The barnacle, of which the swimming larva (nauplius) differs but little from those of other crustaceans, in its host becomes in effect a sort of filamentous bush, totally unrecognizable and evoking movie aliens. A whole group of barnacles of various genera and species parasitize nearly all decapod crustaceans.

The life cycle of this barnacle is as follows: the parasite injects itself through the cuticle of a crab that has just molted. Initially a small cellular mass, it proliferates, branches, and spreads throughout the circulatory system (hemocoel) of the host, becoming an organ that absorbs nutrients. A type of ball, called the *externa,* then projects under the crab abdomen and contains the genitals, then the eggs of the parasite. It occupies exactly the spot in which the eggs of the female crab are normally found. The externa contains an ovary and several testicles. The testicles are in fact tiny males that have grafted onto the externa and then become simply the spermatocyte cells that produce the spermatozoa (this peculiarity caused these

82   barnacles to be considered hermaphrodites until 1958). All *S. carcini* females are crab parasites, and all *S. carcini* males are parasites of *S. carcini* females. The grafting of the males is indispensable to the survival of the female: one with an externa that remains a "virgin" dies after a few weeks.

The manipulation is spectacular. (1) Whether the host crab is male or female, it is completely sterilized. (2) If the host crab is a male, it is feminized. The feminization of males is seen in the morphology and behavior. Morphologically, the abdomen, normally narrow in male crabs, becomes wide, as in females. This widening allows the parasite externa to be sheltered without compression. Furthermore, the appendages located under the abdomen (pleopods) take the normal shape of those found in females. Behaviorally, the feminized male crab acts exactly like a female. It continually "ventilates" the externa, mimicking a female crab ventilating its egg mass. At the moment of eclosion, it assists the dispersal of the barnacle larvae by moving to a zone with strong water currents.

The process of feminization follows the degeneration of the androgen gland in the male crab. However, the manipulative mechanisms are far from being understood in detail. For instance, we do not know why the androgen gland degenerates.

The Case of Theileria

Theileria species are unicellular parasites, transmitted by tick bites, that live in vertebrate blood cells. For example, *Theileria parva* causes East Coast fever of livestock in Africa.

When a tick vector (a species of *Rhipicephalus*) bites a vertebrate, it injects along with its saliva the parasite stage known as a *sporozoite*. The sporozoites penetrate lymphocytes, where they transform into schizonts, in which cell divisions occur—a clonal multiplication that gives rise to new sporozoites that enter other lymphocytes, and so forth. The lymphocytes normally multiply when they recognize a foreign entity, or antigen. In the absence of the antigen that each lymphocyte is programmed to recognize, this multiplication does not occur and the cells are in a resting state.

Once infected by *Theileria*, lymphocytes engage in clonal multiplication, even in the absence of the antigen they would normally recognize as a cue to divide. Instead, the parasite produces a molecular signal that triggers the cascade of events leading to lymphocyte division. The parasites divide concurrently with the lymphocytes. The *Theileria* nuclei formed in the course of this division are even incorporated in the mitotic spindle of the dividing lymphocyte, separating into two approximately equal groups and being drawn along with the lymphocyte chromosomes toward the poles of the spindle. The two daughter lymphocytes are therefore infected, and the entire process is quickly repeated.

This manipulation allows *Theileria* to proliferate its habitat indefinitely, all the while avoiding circulating in the plasma where they could be recognized and attacked by antibodies. The invasion quickly takes the form of a population explosion, and the malady can be lethal.

Careful Use of the Host
Conflict between Two Genomes; or, What Is Virulence?

For a parasite, the host is the goose that lays the golden eggs. The Belgian parasitologist P. J. Beneden was using this expression by the end of the nineteenth century. We deduce from this metaphor, sometimes a bit too hastily, that parasites should be as benign as possible.

I will first define *virulence*. From the perspective of the host, which is the viewpoint taken by humans, we inevitably end up focusing on the effect of the parasitism. Physicians measure pathogenic effect or pathogenicity. By contrast, ecologists and evolutionists measure virulence. Pathogenicity and virulence are in no way equivalent. The notion of pathogenicity encompasses the sum total of the consequences of the presence of the foreign species in the host species. It is determined by observation of particular traits of the host individual, such as morphology, anatomy, metabolism, behavior, reproductive activity, and so forth. For example, if a cow loses weight following infection by the liver fluke *Fasciola hepatica*, this loss constitutes a pathogenic effect. Similarly, if a sheep carrying a cysticercus of the tapeworm *Multiceps multiceps* in its brain begins walking in circles, this behavioral alteration is also a pathogenic effect.

The notion of virulence applies only to the consequences of the presence of the foreign species on the transmission of the host genes. For any sort of pathogen, ecologists designate by the term *virulence* the diminution of host reproductive success it causes.[24] The more virulent a pathogen, the less abundant the offspring of its host and therefore the lower the transmission rate of its genes to the next generation. Any living organism that uses another living organism simultaneously as a resource and as a habitat is pathogenic if it causes a change in the anatomy, physiology, or behavior of its host. It is virulent if and only if it reduces host reproductive success.

If the definition of virulence is straightforward, measuring it is not easy (Poulin and Combes 1999). This is why virulence is too often measured either by just one of its components, such as host mortality (Ebert and Mangin 1997), or indirectly by pathogenic effect, under the assumption that virulence is proportional to pathogenicity (which is obviously far from

84   true). The essential difference is that pathogenicity is important at the level of the individual host (for example, a sick person), whereas virulence is evolutionarily important to the population or species.

Optimal Virulence

We assume that every parasitic organism inhabiting another organism and modifying its resources, even if only slightly, imposes a cost on its host. When the diversion of resources is direct (that is, when the parasite feeds itself at the expense of the host or takes a part of the host's food), whatever is diverted is no longer available to the host. When there is no direct diversion, the cost may be very slight, but it always exists. Virulence—that is, the lowered fecundity of the host—results directly from the loss of resources to the parasite and indirectly from secondary effects, for example, the fact that the host must eliminate (at some cost) the toxic waste products produced by the parasite.

Parasites are selected, as are all living beings, to reproduce as much as possible and thus to exploit their environment as much as is possible. This does not mean that the maximum possible exploitation by the parasite of the host will also be the most favorable to the reproductive success of the parasite. This is because lowered reproductive success of the host (caused, for example, by high mortality) can act against the parasite. The optimal virulence (that which maximizes transmission of the parasite) is therefore not always the maximum virulence.

I consider two examples, myxomatosis in rabbits and chestnut blight.

Myxomatosis is caused by the *Myxoma* virus. This virus is native to the Americas and causes only a benign malady in its original New World hosts. It is transmitted by insect bites (of fleas or mosquitoes, depending on the site). Myxomatosis was introduced during the 1950s to many regions of the world to control rabbit overpopulation. Initially, these introductions achieved their objective perfectly. For instance, in Europe myxomatosis is believed to have killed more than 99 percent of the rabbit population (Kerr and Best 1998). However, with the passage of time, rabbit populations have been less and less affected, although the disease rages on.

What is the origin of this change in virulence? It is certainly partly the result of an increase in resistance among the rabbits—those that manage to survive transmit to their descendants the genes that limited the viral impact. However, comparison of frozen stocks of the virus collected several decades apart shows that the virus nowadays is, on average, much less virulent than the virus that began the epizootic. The explanation of this decrease in virulence is almost certainly as follows (Levin 1996). The most virulent variants of the virus invaded rabbits so rapidly that, statistically speaking, insect vectors were unlikely to be able to transmit them to new hosts before the originally infected hosts died. Therefore, among the diversity of genotypes in the virus population, the most virulent ones were selected against.

We cannot deduce from this reasoning that virulence can decrease indefinitely. In fact, if the virus multiplies too slowly, the probability that the fleas ingest virions in the course of their blood meals declines, which means that the viral genotypes whose virulence is too weak will not be propagated. We can then predict logically that natural selection should maintain virulence at a compromise, intermediate level at which the viral invasion rate suffices for the fleas to become contaminated while biting but not for the rabbits to die too quickly.

Chestnut blight is caused by the fungus *Cryphonectria parasitica*, which is transmitted from tree to tree by various plant-eating insects. There are virulent and hypovirulent (less virulent) strains of the parasite. Curiously, the weakly virulent strains are caused by a parasite of the fungus, a DNA parasite of the type discussed in chapter 2 (Taylor et al. 1998). If the two strains compete on the same tree, the virulent strains outcompete and eliminate the hypovirulent strains, and the host tree dies quickly. This competitive outcome causes a mass chestnut death of the sort seen in North American and European forests in the first half of the twentieth century.

When chestnuts become too rare, however, the advantage shifts. In effect, the less virulent fungi become increasingly favored because the fact that they do not kill the tree too quickly means that the probability that insects transmit them to a new tree is greater than for the more virulent strains (Newhouse 1990).

It is evident that, as in the myxomatosis example, virulence cannot surpass a certain optimal level above which the pathogen finds its transmission rate compromised (Ebert and Herre 1996), thereby decreasing its reproductive success. Therefore, there exist potential situations in which virulence is below the optimum (parasite transmission would be augmented if virulence were greater) and situations in which virulence is above the optimum (transmission would be augmented if virulence decreased).[25]

We can conclude that, for the parasite, except in special situations, virulence is a two-edged sword: if virulence decreases host reproductive success too much, it will produce a reaction in the form of selection for lower virulence (Combes 2000, 2001).

What Determines Optimal Virulence?

There is no general answer to the question What determines optimal virulence? because the answer differs for each association of parasite, host, and environment. We can at best deduce some rough rules. I examine two of them, one about horizontal or vertical transmission of a parasite and the other about competition between parasites.

We call any infection that occurs between individuals of the same population (whether related or not) *horizontal* transmission. By contrast, any system in which the pathogen is transmitted systematically from host generation n to host generation $n + 1$ is called *vertical* transmission. When transmission is entirely vertical, the only individuals the pathogen can infect are descendants of the infected host individual. We can guess that, in vertical transmission, parasite reproductive success is linked to host reproductive success. The reaction by natural selection to virulence is maximized, and we can predict that this selection will reduce virulence to the minimum possible level compatible with exploitation of the host. The result should be the weakest possible virulence.

Herre (1993, 1995) has shown the relationship between vertical transmission and reduced virulence in research that has become a standard reference on nematode parasites of wasps that pollinate figs. These are very specific associations: each wasp species pollinates one fig species. (I discuss these remarkable associations again, in the context of mutualism, in chap. 5.)

When a female wasp penetrates a fig, she lays her eggs, then dies. If she is parasitized by nematodes of the genus *Parasitodiplogaster,* these also lay their eggs in the fig, then die. Later, still in the fig, nematode larvae parasitize young wasps that result from the development of wasp eggs. The number of female wasps that penetrate the same fig differs depending on species; in some species, there is but a single founder female, whereas in others there are several. What is interesting about this situation?

If a single wasp enters the fig and is parasitized, the reproductive success of the nematode depends on that of the wasp: transmission is strictly vertical. If the nematode is too pathogenic (this pathogenic effect occurs before the penetration of the fig), the wasp that houses it will lay fewer eggs, so descendants of the nematode will have few young wasps to infect. If several wasps penetrate the same fig and if some of them are infected, the reproductive success of a particular nematode does not depend solely on that of the particular wasp that houses it, because the fig will also have offspring of other wasps. Transmission is at least partly horizontal (fig. 3.15).

Herre's prediction was that virulence should be lower in the first case (vertical transmission) than in the second (horizontal transmission). By

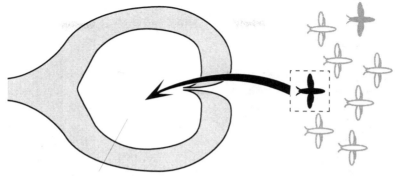

The fig will harbor:

offspring of the nematode
+
offspring of the infected wasp

A

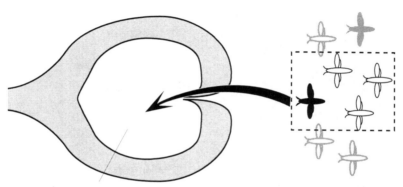

The fig will harbor:

offspring of the nematode
+
offspring of the infected wasp
+
offspring of uninfected wasps

B

Fig. 3.15. Vertical (A) and horizontal (B) transmission. When *Parasitodiplogaster* parasitizes a wasp species of which a single female enters a fig (A), transmission is strictly vertical (the nematode larvae can infect only descendants of that female). When *Parasitodiplogaster* parasitizes a wasp species of which several females enter a fig (B), transmission is partly horizontal (the nematode larvae can infect descendants of unparasitized wasps). Virulence is stronger in case B than in case A. Parasitized wasps in black; frame encloses wasps that enter the fig.

88 counting both the founder wasps, whose corpses are found in the fig, and the young wasps, Herre was able to determine the decrease in fecundity (a measure of the virulence) caused by the parasites. The results showed that the prediction was completely accurate. Vertical transmission was indeed accompanied by reduced virulence.

The profession of parasite therefore consists of being as virulent as possible, but it must be understood that the "possible" can range from very limited to unlimited. In some cases, for example, when transmission implies the death of the host, virulence is given free rein. In other cases, because the parasite exploits a fragile, living environment that is an absolute necessity, natural selection protects the goose with the golden eggs.

An important reason why virulence of a parasite in relation to its host may not be optimal is competition between parasites. In effect, when there is strong competition within the same host, either between several parasite species (interspecific competition) or between genetically different individuals of the same species (intraspecific competition), selection favors the most active competitors (for example, those that reproduce most quickly), and virulence increases (see Read and Taylor 2001).

Coluzzi (1999) appeals to competition to explain the high virulence of one malarial agent, *Plasmodium falciparum.*

Currently, *P. falciparum* causes the gravest cases of malaria and infects only humans. However, research on DNA sequences shows that *P. falciparum* has a close relative, *P. reichenovi,* that infects chimpanzees and gorillas. It appears that the divergence between *P. falciparum* and *P. reichenovi* dates to 7 million years ago, which corresponds to the date usually assigned to the separation between "apes" and hominids. Speciation occurred simultaneously between primate hosts and between their parasites.

An ancient association between humans and *P. falciparum* fits poorly with the observed strong virulence and short duration of the infection in humans, however. These traits are disadvantageous for parasite transmission and should logically have been selected against. Coluzzi believes that, at a time when the human population was limited to small hunter–gatherer groups, such traits would have been even more likely to drive *P. falciparum* to extinction.

Coluzzi therefore imagines that the strong virulence of *P. falciparum* for modern humans is a recently acquired trait. He observes that, in the Neolithic period, about 6,000 to 8,000 years ago, humans caused two major changes in the ecology of *Anopheles gambiae,* the most important mosquito vector of *P. falciparum.* First, humans created very favorable egg-laying sites for the mosquito by clearing and exposing to strong sunlight patches of

The Profession of Parasite

moist forest receiving heavy rainfall. Second, the establishment of sedentary human communities (villages) led *A. gambiae* to specialize in biting
humans. These two changes greatly increased the aperture of the encounter
filter between *P. falciparum* and humans.

Opening this encounter filter then induced a real "saturation" in human
infection: even today, many inhabitants in the African malaria zone receive
more than a hundred infective bites (that is, by mosquitoes carrying *P. falciparum*) every year. Such saturation can only augment the intraspecific
competition between various *P. falciparum* genotypes. Now, as I have
pointed out, such competition nearly always entrains increased parasite
virulence. Coluzzi believes this is the explanation of the present-day strong
virulence of *P. falciparum*. The key changes in the mosquito demography
are thought to have occurred just a few millennia ago. In fact, all *P. falciparum* today could conceivably have descended from a single very recent
mutant individual (Volkman et al. 2001).

Of course, such increased virulence should cause selection for resistance
traits in human populations, and such evolution is observed today. For
example, locally there is an increased frequency of a gene coding for a
hemoglobin variety (hemoglobin C; see Modiano et al. 2001) that hinders
Plasmodium growth, but selective processes in humans are slow, and today
they are still far from having released humans from the danger of malaria
caused by *P. falciparum*. The situation can be even more complex. For
instance, certain bacteria may include highly pathogenic genotypes that
rapidly build up populations in the host but are not adept at transmission,
whereas other, less pathogenic genotypes comprise the majority of transmitted individuals (Dye 2001).

A Parasite Should Leave Its Host

From One Island to Another

Parasites live, as I have said, as if on islands separated from one another
by an ocean of inhospitable habitat. In this metaphor, islands represent
host individuals, and the ocean represents the external environment a parasite must brave to pass from one host to another. Every parasite must
therefore have adaptations allowing it to leave its host. I should note that
this is just a particular case of a process common to the life cycles of all living organisms—dispersal.

The image of a dandelion casting to the wind its seeds suspended on delicate parachutes is a classic illustration of dispersal. In fact, reproduction in
animals as in plants is always followed by a stage during which "propagules"

generated by reproduction leave the environment in which they were born to try to master a new environment similar (at least in its major features) to the one they have left.

The parasite version of this process is peculiar only in that, when the parasite inhabits the interior of its host (for example, the host circulatory system), exiting it requires a sort of escape.

Story of a Host Death Foretold

The life cycle of a trematode, *Aphalloides coelomicola,* shows us the simplest basic solution. We then see that it is rarely applied in nature.

Claude Maillard (1976) discovered the transmission mode of this fish parasite in Mediterranean pools. At its beginning, this cycle is a classic one for a trematode: a prosobranch mollusc (*Hydrobia ventrosa*) ingests the egg containing the miracidium. The parasite proliferates in the mollusc genital gland, producing rediae from which the cercariae escape. The cercariae leave the mollusc and, swimming poorly, remain on the bottom of the pool waiting for small gobies (*Potamoschistus microps*) to pass by. If contact with a goby occurs, the cercariae penetrate the integument and encyst as metacercariae, mainly in the wall of the body cavity (coelom). After ten days, the metacercariae emerge from their cysts and fall into the fish body cavity. There the parasites mature, rapidly becoming sacs stuffed with eggs that accumulate (up to 90 or more) in the coelom. No regulation appears to exist; the more numerous the cercariae that enter, the more numerous the adults that invade the body cavity.

Maillard quickly recognized that part of this life cycle is highly unusual for a trematode. Normally we expect the metacercariae to enter a quiescent stage in the goby and the parasite to produce eggs only in a third host that eats the goby. There has therefore occurred a sort of acceleration of the classical trematode life cycle: instead of "waiting for" the third host, the parasite matures and produces eggs in the second; the expected third host does not exist. Maillard's hypothesis is that the eggs are liberated when the goby dies, whether naturally or by being eaten by a predator.

More than twenty years later, C. Pampoulie et al. (2000) again took up research on *A. coelomicola* and showed that Maillard's hypothesis was exactly right: the parasite propagules (that is, the eggs) accumulate in the coelom and leave the goby at its death. How and why did this exit mode evolve?

The explanation lies in the fact that the goby *Potamoschistus microps* lives only one year. Juveniles appear in pools in June, grow until the following spring, reach sexual maturity, reproduce, then die. There thus exists a yearly cohort with a completely predictable fate. The parasite infects fish beginning in autumn and accumulates eggs until the following spring. At the beginning of the summer, the death of the gobies releases the parasite eggs into the water.

We now see the advantages the parasite draws from maturing wholly within the goby. On the one hand it avoids the vagaries of transmission to a third host, and on the other it profits from the programmed death of its host to spread itself throughout the pool at a time of the year when the molluscs it is going to infect are themselves available. We can add that the goby body cavity probably provides a rich and protected environment. Any use of a third host could only slow and perhaps halt successful transmission.

A. coelomicola therefore illustrates a remarkably simple exit mode for a parasite: host death, accompanied by rapid decomposition of its tissues. However, we can guess that only the peculiar biology of gobies allows selection for this mode of exiting the host. Although some other parasites use this same method, this adaptation is unusual. When the timing of host death is not so predictable, or when there is a risk that it will occur outside the habitat in which the life cycle unfolds, other exit strategies should be found.

Ectoparasites (living on the surface of their hosts) and mesoparasites (living in host cavities that communicate with the outside) expel their propagules like sessile free-living animals. There is no barrier between the parasite producing the propagules and the environment in which these disperse. The situation is very different for endoparasites, separated from the outside by one or several barriers.

"Autonomous" Exit

In their adult stage, schistosomes are parasites of birds and mammals. Their life cycle is similar in some respects to that of *Aphalloides coelomicola* in the sense that there are only two hosts, maturity is attained in the second host, and the adult is an endoparasite.

Let me summarize the *Schistosoma* life cycle: the egg hatches into a swimming miracidium that penetrates a freshwater gastropod mollusc. Multiplication within the mollusc gives rise to sporocysts, then to swimming cercariae; these penetrate the integument of the definitive (final) vertebrate host and reach the blood vessels. Whereas almost all trematodes are hermaphrodites, schistosomes have separate males and females. Copulation takes place in the blood vessels of the host liver, and permanent pairs are formed, the male carrying the female in the folded edges of its ventral surface. These pairs do not settle just anywhere. In intestinal schistosomiasis (caused by *S. mansoni* and *S. japonicum*), the adult parasites lay their eggs in the blood vessels adjacent to the intestine. In urinary schisto-somiasis (caused by *S. haematobium*), the parasites deposit their eggs in blood vessels of the bladder wall. Certain parasites of bovids (*S. nasale* in India) and birds (*Trichobilharzia regenti* in Europe) inhabit the blood vessels of nasal fossae and lay their eggs there.

Although these parasites are prisoners in the circulatory system, we see that their locations are not random; they are always situated near an "exit": the intestine, the urinary tract, or the nose. There still remains the most difficult feat, to cross the tissue barrier that separates the eggs from the natural exit near which they were laid. Three kinds of adaptations, very different but complementing one another, allow schistosome eggs to escape their host.

The eggshell in most species of schistosomes has a pointed spur, the adaptive value of which is not difficult to guess: it allows the rending of host tissues in the immediate vicinity of the egg by virtue of the natural movements of the organs of the infected individual, for example, intestinal peristalsis.[26] An electron microscope examination of the schistosome eggshell shows it to be riddled with a plethora of tiny orifices. The main function of these orifices is seen while the egg traverses the tissues: they allow the passage from the egg of proteolytic enzymes that induce necrosis in the surrounding tissues. This necrosis helps the egg to reach the intestinal lumen and aids its mixing with contents of the intestine, the bladder, and the nasal fossae.

The most remarkable adaptation, however, is the "use" the parasite makes of the host defense mechanisms. In response to a foreign object such as a schistosome egg, a vertebrate brings into play the immune defenses, which are primarily cellular. A capsule (granuloma) forms around the egg in reaction to its presence; this is a mass of cells (granulocytes, lymphocytes, macrophages) that can grow to be several times larger than the egg itself.

In the late 1970s, several researchers, especially Doenhoff et al. (1978), noticed that the excretion of eggs is less pronounced in immunosuppressed mice than in normal mice. More remarkably, it sufficed to inject blood serum from normal mice into immunosuppressed mice to restore the normal excretion of eggs. These observations indicated that an intact immune system aided schistosome eggs to reach the exterior. Later, it was confirmed that strongly immunodeficient mice excreted almost no eggs, even when the parasite was producing eggs at a normal rate.[27]

Cells of the granuloma are believed to secrete enzymes that help clear the way for expulsion of eggs. The parasite exploits the host immune reaction to its benefit. We say that the granuloma "chaperones" the egg through the host tissues (McKerrow 1997). The schistosome example shows that, with respect to adaptations to exit the host, parasites even use mechanisms that are supposed to fight them.

We can ask why schistosomes do not use the exit path that appears to be the simplest of all—that is, why do they not settle near the skin and lay their eggs in the finest cutaneous blood vessels. Of course, it is not worth attempting to answer that question by elaborating hypotheses about all the advantages and disadvantages of all possible exit routes. On the other hand, it is interesting to note that the adaptation of exiting by the integument has been selected for in other endoparasites—for example, in the human parasite known as the guinea worm, *Dracunculus medinensis*.[28] This disease exists mainly in the African Sahel. Females of this nematode live in subcutaneous tissue, mostly in the lower extremities, and cause ulcerations of the legs that allow them to expel their larvae into the water when the afflicted individual bathes in a pond.

Exit by the Food Chain

Life functions because of energy exchanges, which constitute links in food chains or food webs. The classic food chain comprises producers (plants that capture solar energy by using chlorophyll), primary consumers (herbivores), secondary consumers (carnivores), and decomposers (bacteria).

Imagine a parasite found in a link of this chain—for example, in a field mouse (primary consumer). This rodent is often prey of foxes (secondary consumers). When a fox eats a field mouse, logic would dictate that the parasite carried by that mouse would be digested along with the tissues of its host. However, if special adaptations permit the parasite to survive in the fox, there is a double benefit: the parasite exploits first the rodent, then the fox. There would have to be other adaptations later in the life cycle to allow it to parasitize a field mouse again. The life cycle I have just described is in fact that of *Taenia crassiceps*, a tapeworm that continually cycles from field mouse to fox and from fox to field mouse (but never from field mouse to field mouse or fox to fox).

This manner of exiting a host by entering another host in the course of a predatory act is very widespread among parasites.

In general, when parasite transmission is by the food chain, the host from which the parasite comes (called the upstream host) and the host it enters (called the downstream host) belong to different species (fig. 3.16). However, I must mention the exceptional case of the trichina (*Trichinella spiralis*), which can be transmitted indefinitely by predation within one host species (rats, as described earlier in this chapter).

Fig. 3.16. Parasite transmission via a food chain. (A) The classic case (passage from a primary consumer to a secondary consumer); (B) the trichina case (indefinitely prolonged circulation by cannibalism within one population).

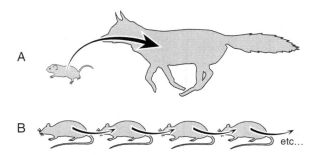

We also know that some parasites profit from the existence of biting insects that, from blood meal to blood meal, travel among potential hosts of the parasite, thus spending most of their time crossing frontiers for the latter.

Controversy: Adaptationism and Parasites

The Spandrels of the Basilica of Venice

No one denies that every extant species is adapted to its environment. By contrast, the idea that every structure, every process, or every behavior necessarily results from natural selection need not be true. It was S. J. Gould and R. C. Lewontin (1979, 1982) who, if one can use this metaphor, sounded the alarm. In a famous article titled "The Spandrels of San Marco and the Panglossian Paradigm: A Critique of the Adaptationist Programme," these authors developed the following idea.

Ever since biologists have begun to understand the process of adaptation, they have had a natural tendency to assume that all "traits" of living beings, whether morphological, physiological, or behavioral, result from natural selection. This is equivalent to saying that all traits are adaptations for something or other. This is the view that Gould and Lewontin call the *adaptationist program.*

In one sense, the adaptationist program is another version, albeit with a different mechanism, of the ancient idea that the world, created by an infallible God, is perfect in every detail. This idea was propounded by, among others, Bernardin de Saint-Pierre,[29] who believed that oranges have their striking color in order that we can easily see them in trees and that a melon has sections so that it can be shared in an equitable fashion by family members.[30] Doctor Pangloss, Candide's tutor in Voltaire's famous pamphlet, has come to symbolize this attitude, endlessly proclaiming that "all is for the best in this best of all possible worlds."

To demonstrate that every trait need not reflect a biological adaptation, Gould and Lewontin have used an original metaphor. Look, they say, at the mosaic decoration of the great central dome of the Basilica of San Marco in Venice: three circles of figures radiating from the central representation of Christ: the angels, the disciples, and the virtues. Each circle is itself divided into quadrants, and each quadrant joins one of the four spandrels that support the dome—that is, the subtriangular zones that are formed by the intersection at right angles of two rounded arches (fig. 3.17).

Gould and Lewontin (1979, 581) note two apparently contradictory facts: (1) each spandrel is decorated with a figure of great artistic quality that fits perfectly in the subtriangular frame[31] ("each spandrel contains a design admirably fitted into its tapering space"); and (2) the spandrels constitute the only architectural structure that allows the construction of a dome on circular arcs ("are necessary architectural by-products of mounting a dome on rounded arches").

The risk arises immediately that the spandrel decorations may be considered the raison d'être of their shape and that this decoration is thus the "cause" of the entire architectural structure of the Venetian dome. In fact, the structure is, on the contrary, the result of an architectural constraint. It is the artists who adapted their work to the architectural structure, and not

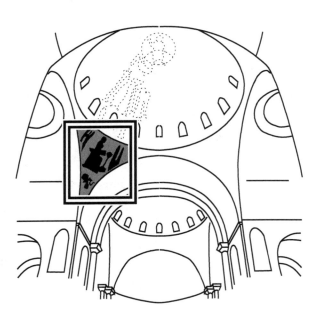

Fig. 3.17. The spandrels of the dome of the Basilica of San Marco in Venice, famous among biologists thanks to S. J. Gould and R. C. Lewontin. A single spandrel is detailed.

96 the reverse. The spandrels are "by-products" of the dome and not adaptations to the decorations, however successful the latter are.

If one transposes this spandrel metaphor to biology, the warning is clear: when a particular trait of an organism appears well adapted to its use, one must not, to use the phrase of Massimo Pigliucci and Jonathan Kaplan (2000, 66), "jump to the conclusion that the particular use is the reason the trait is present." There may be other reasons for the origin of the trait than the adaptation we see today.

The Work of Doctor Pangloss

Parasitism brings strong arguments to this debate, both because of the omnipresence of adaptations in the living world and because there are constraints that may prevent the traits of organisms from being the best of all possible ones in this best of all possible worlds. Parasite life cycles demonstrate both the strength and the weakness of adaptationism.

Let us consider in more detail than we did earlier the life cycle of trematodes in the genus *Halipegus*. Recall that there are four obligate hosts, as shown by Nadia Kechemir (1978). In Europe:

- The cercariae of the species *Halipegus ovocaudatus* are produced in a mollusc of the genus *Planorbis* (first host).
- The cercariae, released in water, are shaped like little spheres having a motion detector and a syringe; when grabbed by a copepod (second host), they inject themselves into its body cavity in a fraction of a second.
- When a dragonfly larva (third host) eats the copepod, the parasite lodges in the larval digestive tract.
- If a green frog (*Rana ridibunda*) (fourth host) eats the dragonfly, the parasite settles in the buccal cavity, under the tongue, and lays its eggs there.

All these stages are obligatory.

Dr. Pangloss's attitude, had he known this prodigiously complicated life cycle, would surely have been to consider that what *Halipegus* does is in every way the best thing it can do to survive and make its way on this planet.

First remark: Dr. Pangloss would certainly have been correct to marvel at the adaptations incorporated in this life cycle—for example, the mechanism that injects the cercaria into the copepod is a little engineering marvel. Similarly, the fact that the parasite can survive both when the dragonfly eats the copepod and when the frog eats the dragonfly implies that an entire molecular arsenal has been selected that allows this survival.

Second remark: Dr. Pangloss would have been only partly correct, for the following reason. *Halipegus* belongs to a trematode family, the Hemiuridae, whose species typically have a life cycle with three hosts: a mollusc, a copepod, and a fish. In fact, *Halipegus* is the only genus in this family that has a frog as the definitive host rather than a fish. It is therefore reasonable to assume that the life cycle with a fish as the definitive host is the primitive state and that the cycle with the frog as definitive host results from more recent evolution.

Let us imagine an engineer (we could even name him Pangloss) who is tasked with adapting the primitive life cycle to the conquest of a new type of host, an amphibian. We immediately realize that he ought to take into account the facts that (1) frogs are terrestrial and do not eat copepods (neither do the tadpoles, which are herbivores); and (2) frogs eat dragonflies, and dragonflies have aquatic larvae. He would then construct a three-host life cycle, mollusc–dragonfly–amphibian, in which the dragonfly would simply have replaced the copepod: one arthropod would have been traded for another arthropod, and the substitution would seem easy. The engineer would also consider that other trematodes have this type of life cycle and that, after all, what already works in other species should be a good indicator of success for *Halipegus*.

Alas, evolution did not listen to Engineer Pangloss (fig. 3.18). The parasite was obviously unable to abandon its ancestral attraction for copepods, and this fact is the basis of the solution of using the dragonfly as an extra intermediary connecting the copepod and the frog. This solution was adopted at the price of having one extra transmission episode, and therefore a probable extra loss of infective stages.[32] This is another example showing that, when shaping life cycles, evolution sometimes acts more like a tinkerer than an engineer.

The logical conclusion is that even if the life cycle of *H. ovocaudatus* functions well as it is, it is equivalent to the spandrels of the basilica of Venice: it inspires admiration,[33] but it is the result of a constraint, namely, the architecture of the ancestral life cycle of the Hemiuridae.

It is highly likely that multiple constraints (we speak of phyletic constraints and developmental constraints) impinge on the morphology, physiology, and transmission modes of many parasites. Parasites marvelously illustrate the conclusion of Pigliucci and Kaplan (2000, 66): "It is the synthesis of constraints (spandrelism) and selection (panglossianism) that is the key to a more sober and realistic understanding of phenotypic

Fig. 3.18. A group of parasites conquers amphibians as hosts, by a tortuous application of natural selection. (A) The ancestral life cycle of Hemiuridae; (B) the cycle that would logically have led the parasite to amphibians; (C) the solution produced by evolution, using a dragonfly as an extra intermediate host.

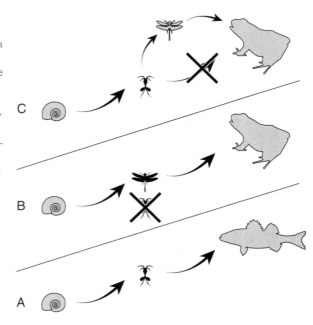

evolution. . . . Biology is partly an experimental science, but partly a historical one."[34]

Because they have frequently changed lifestyle, if only to pass from a free-living state to a parasitic one, parasites have many traits that also illustrate the concept of *exaptation,* defined by Gould and Vrba (1982) as designating any trait whose present function has nothing to do with the function that was the basis of its selection. Pierre-Henri Gouyon (1998, 43) observes, for instance, that "the eye is made to see—natural selection has favored this possibility—but the tongue, a precondition for language, did not arise evolutionarily so we could speak." Language thus results from an exaptation. Of course the concept of exaptation does not imply that the organ that acquires a new function is not subsequently modified by natural selection so as to fulfill its new function as well as possible. Exaptations are surely numerous among parasites.

What Good Are Vectors?

In several parts of this chapter we have encountered the concept of vector. I have applied this term to many species that harbor trematode metacercariae, to biting insects that transmit viruses, protist parasites, or filarial worms. I could multiply the examples. In introducing the controversy over adaptationism, I want to mention an adaptation to transmission that is possibly the most extraordinary, or at least the most worthy of reflection when one hesitates between attitudes that are too "Panglossian" or too "spandrelian." A vector of unexpected identity is implicated in this case.

I have said that, to detect their target hosts, insect parasitoids deploy an elaborate arsenal of sensory structures that allows them first to find plants where they "know" they are likely to find their potential victims and then to identify the victims themselves, which are larvae of other insects. It is the adult female parasitoids that conduct this search and lay their eggs on the "prey." The parasitoid larvae then develop at the expense of the host insect, which ultimately dies in almost all cases.

With respect to discovering its target, the little wasp *Mantibaria mantis* proceeds differently. Larvae of this parasitoid develop at the expense of the praying mantis *Mantis religiosa*. I quote the description of its strategy by Bernard Chaubet (1996, 106):[35]

> *Mantibaria mantis* . . . sets off to find an adult mantis and settles under the wings of the giant insect. This clandestine passenger than begins a sedentary life that will last several months. . . . [When the mantis is about to lay eggs], the wasp descends the length of its host and reaches the material surrounding the eggs. . . . At risk of being gobbled up, it drags its abdomen until it discovers a mantis egg, then deposits its own. The *Mantibaria* larva develops to adulthood at the expense of the mantis egg.

The strategy is unstoppable; the parasitoid travels hidden in the wings of its adult host as long as is necessary while waiting for the host to reproduce; the wasp doesn't have to seek the host's eggs because it is sequestered just above them.

The mantis is the vector of its own parasite. Bernardin de Saint-Pierre would have been flabbergasted.

And the Manipulations?

The sort of question I have just posed on the adaptive value of various parasite traits is particularly apt for the extended phenotype (see the definition in chap. 1 and the examples of manipulation in this chapter). Very many parasite species, in diverse ways and to varying degrees, change the phenotype of their hosts, and it is therefore tempting for a parasitologist to attribute an adaptive value to these changes. There is no dearth of examples in which no experiment is possible, so the hypothesis of the adaptive extended phenotype is untestable.

I cite two examples. Is the fact that we sneeze when we have a cold the result of a strategy selected for in the virus because sneezing favors the transmission of the cold to nearby individuals? Is the fact that pinworms in children, by laying their eggs, cause the anus to itch the result of selection because the itching induces scratching and scratching aids dissemination of eggs (the eggs stick to the fingernails, then to the toast and jam, thus re-infecting the child or one of her playmates)?

We should also recall (last chapter) that, in the spontaneous evisceration of sea cucumbers, it is possible to see an adaptation of the host that permits it to rid itself of parasites, or an adaptation of the parasites that assures their propagules are disseminated, or, of course, a physiological reaction of the host that has nothing to do with parasitism.

As Robert Poulin has often emphasized (1992a, 1995, 1998), the simplest explanation for the observed changes (for example, the poorly joined shells of the cockle or the aberrant behavior of ants) is that they are pathogenic effects among many that are possible. In some cases, the phenotypic modification, even if spectacular, is certainly not a result of manipulation. At the same time, it is intellectually unsatisfying to reject the idea that natural selection has generated some of these changes, maybe many of them. We are again at the heart of the debate between Panglossism and spandrelism.

When the favorization process is manifested in transmission by a vector's eating a parasite infective stage, it is possible to undertake comparative predation tests of the sort I described earlier with respect to transmission of the trematode *Microphallus papillorobustus*. These tests consist of offering the predator normal individuals and individuals with parasite-modified behavior.

Such research is very difficult, especially if one wants to undertake it not in the laboratory but in nature (which is in principle the only approach giving unassailable results). To my knowledge, no one has offered samples of parasitized cockles and healthy ones to oystercatchers, and no one has been able to count the number of healthy ants and the number of parasitized ones in sheep stomachs and compare them to the proportions available in pastures. Fortunately, there are a few life cycles that have allowed such comparative research.

Several parasites suspected of manipulating their hosts are found in host nerve centers. This is the case for *Euhaplorchis californiensis*, a trematode studied by Kevin Lafferty and Kimo Morris (1996). Its metacercariae are found in great numbers in the brain case of a small marine fish, *Fundulus parvipennis* (6–7 cm long). The definitive host is a seabird that becomes parasitized by eating this fish.

The question is, Why are the parasites found in this particular place in the fish? Has the location been selected to escape the immune defenses, or has it been selected because it increases the probability that a bird will capture the fish? Lafferty and Morris first noticed in the aquarium that the behavior of parasitized fishes is characterized by disordered movements not seen in healthy fishes. At this point, the parasite could be suspected of manipulation but could not be formally charged. This is why they undertook field tests in fish pens delimited by nets along the shore at Santa Barbara, California. The pens were seeded with healthy fishes and parasitized ones in known numbers, and seabirds could easily come and forage in one of these pens. A pen from which birds were excluded served as a control. The fishes were recaptured after 20 days in each of the two pens, and whether they were parasitized by *E. californiensis* was noted.[36]

Results are shown in figure 3.19. In the pen from which birds were excluded, the number of healthy fish declined from 53 to 50 and the number of parasitized fish from 95 to 91. In the pen exposed to birds, the number of healthy fish fell from 53 to 49 and the number of parasitized fish from 95 to 44. By computer simulation, Lafferty and Morris showed that the parasitized fish are 30 times more susceptible to predation than are healthy fish. Lafferty and Morris also counted metacercariae of *E. californiensis* in the brain cases of parasitized fish; the number

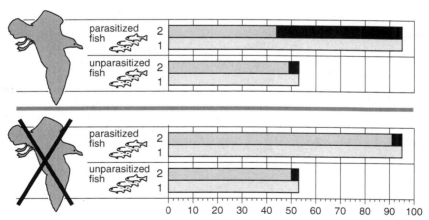

Fig. 3.19. Test of facilitation of predation. Fishes parasitized by *Euhaplorchis californiensis* are approximately 30 times more likely than healthy fish to be eaten by seabirds. The black band represents fishes missing (= eaten) at the end of the experiment. I = beginning of experiment; 2 = end of experiment.

varied from 500 to 3,000, and comparison between the two pens showed that fishes with the heaviest parasite load were most likely to be eaten.

The results of these experiments obviously do not prove that all host behavior modified by parasites has an adaptive value for transmission. But at least these results are clear and cannot be otherwise rationalized, and they concern real species in nature, not a parable about the spandrels of a Venetian basilica.

The Profession of Host

The Two Lines of Defense

As I have shown in the two notions of encounter filter and compatibility filter (chap. 1), hosts theoretically have two lines of defense. The first consists of avoiding the parasite by appropriate behavior; if the encounter nevertheless takes place, the second consists of attempting to expel the intruder.

In predator–prey systems, the prey are, so to speak, at a disadvantage in that they have but one line of defense, which consists of avoiding the predator. If a small forest rodent meets a hungry fox, its fate is sealed the moment it is between the jaws of the fox (fig. 4.1). Although the prey has no choice in its defensive strategy,[1] the host, by contrast, can "choose" between two strategies. The question I now address is, Do hosts use the first line of defense more frequently, or the second?

The thought that immediately comes to mind is that the answer should differ from host species to host species and might even differ, for a given host species, depending on the parasite species. This surmise is correct, but there nevertheless is a general tendency to use the second line of defense (immunity) more often than the first (behavior). In other words, the profession of host entails more killing than avoiding. Why?

When I discussed the profession of parasite (chap. 3), I pointed out that although hosts that are targets of parasites emit many kinds of information, parasites (except for insect parasitoids) make little use of this information. This was explained by the fact that, even though available information

The genetic diversity as regards resistance to disease is vastly greater than that as regards resistance to predators.

◊

J. B. S. HALDANE (1949)

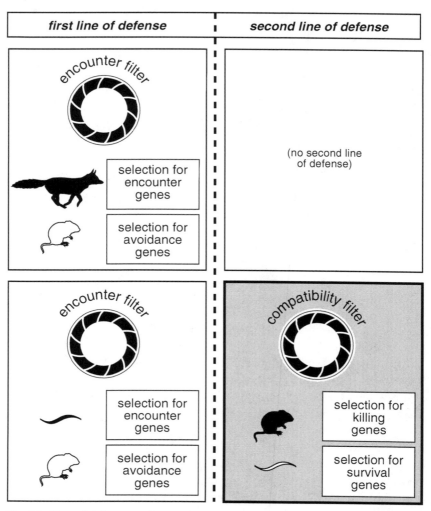

Fig. 4.1. Lines of defense: one for prey (*top*), two for hosts (*bottom*). What happens in the gray square (*lower right*) is important in parasite–host relations; it is here that the persistent interaction is constructed.

abounds, infective stages of parasites are poorly equipped to detect it. Are hosts better able to detect the infective stages of parasites?

The host, target of the parasite, is the analog of the prey, target of the predator. We would therefore expect that host–parasite systems and predator–prey systems would function similarly. But we find nothing of the sort. And this is for a very simple reason—the quantity of information emitted by an infective stage is much less than that emitted by a predator. The infective stage is nearly always small, indeed minuscule, and it is neither smelly nor noisy, which is a great advantage for a parasite (fig. 4.2). Hosts are in a situation nearly the opposite of that of their parasites: they are in principle endowed with very sophisticated means of receiving information, but the information coming from infective stages is so meager that hosts are usually unable to detect them. It is therefore not shocking that investment in the second line of defense (immunity) is greater than investment in the first line of defense (behavior).

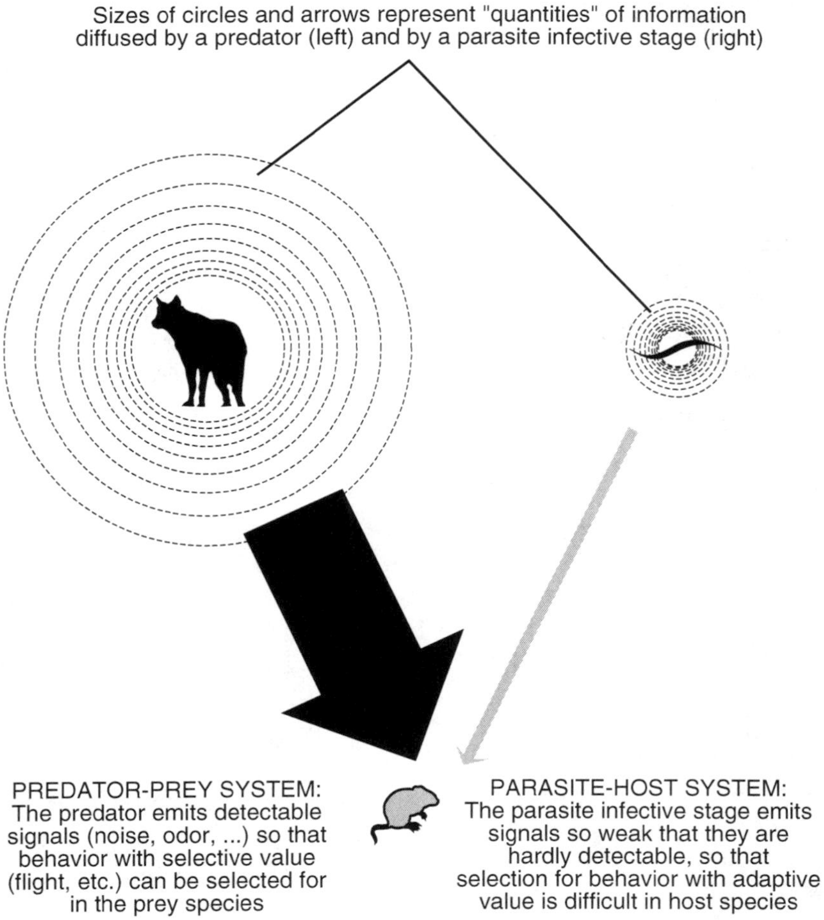

Sizes of circles and arrows represent "quantities" of information
diffused by a predator (left) and by a parasite infective stage (right)

PREDATOR-PREY SYSTEM:
The predator emits detectable signals (noise, odor, ...) so that behavior with selective value (flight, etc.) can be selected for in the prey species

PARASITE-HOST SYSTEM:
The parasite infective stage emits signals so weak that they are hardly detectable, so that selection for behavior with adaptive value is difficult in host species

Fig. 4.2. Although a rodent can detect information (carried by odors or vibrations, for example), it is unable to receive the nearly nonexistent information coming from a parasite infective stage.

Hosts change their behavior to avoid infection in only two cases: when particular circumstances allow detection of the parasite, and when (and this is only in humans) culture allows a deeper understanding of the parasite. I save the special case of humans for the final controversy of this chapter. For now, I discuss an unusual host–parasite system in which the hosts can actually close the encounter filter. The first line of defense, when it can be effective, saves energy that would otherwise be invested in an immune response.

The First Line of Defense

Tits in Switzerland

Heinz Richner and his team have chosen as a study system the tits of the Lausanne and Bern regions of Switzerland. Their research on the relationships between tit populations and their parasites was designed to analyze how parasitism interferes with the lives of the tits and the various means of defense the birds can muster.

The great tit (*Parus major*) is the largest tit of European forests and gardens. Its bluish black head with white cheeks, yellow breast with a black median stripe, and green back are familiar to any European with the slightest interest in birds. The blue tit (*Parus caeruleus*) is slightly smaller. Its white and blue head, blue wings, and yellow breast are also well known.

Tits are insectivores. During the reproductive season, monogamous couples form and nest in natural cavities but also readily accept artificial nestboxes. The time of year when females begin to lay eggs varies with climatic conditions. All evidence suggests, however, that the availability of caterpillars is more important than temperature in triggering egg laying. Caterpillar availability, in turn, is related to the appearance of buds and new leaves on trees. Tit clutch size varies from 4 to 12 eggs.

Tit nests often harbor ectoparasites, fleas of the species *Ceratophyllus gallinae*. These fleas are found on many bird species. However, they are found especially often on forest birds that nest in cavities a certain distance aboveground. They clearly favor the family Paridae—that is, tits (Tripet and Richner 1997a). There are normally two flea generations per year. Fleas of the second generation appear in nests just before the nestlings fledge; some of these fleas leave the nest with the fledglings, whereas others pass the winter in the nests, as cocoons, and it is not uncommon to see hundreds suddenly appear in the spring if a tit settles in a nestbox. When they are very numerous, the fleas, which suck blood from the birds, have a pronounced effect on tit reproductive success.[2] We will see that tits also harbor endoparasites, such as the malarial agent *Plasmodium*.

With respect to the ectoparasites, research by Richner and his colleagues has shown that tits use two means to close the filter of encounter at least partially: they "detect" parasites directly, and they "predict" the presence of parasites and modify clutch size accordingly.

Detecting Parasites

Richner and his team (Oppliger, Richner, and Christe 1994; Heeb et al. 1999) first studied the effect of fleas on the quality and number of young tits by manipulating the number of fleas in nestboxes set in woods around Lausanne and Bern. In this experiment as in those to be described, the nestboxes are of wood, can be easily disinfected, and allow the placement (or not) of a nest of twigs that had built by tits and collected the preceding year.

The experimental nestboxes were divided into two categories—infested and uninfested. Uninfested nestboxes were obtained by a series of treatments in a microwave oven between the moment of egg laying and that of fledging. The infested nestboxes were obtained by three manual contaminations several days apart. For the two nestbox categories, it was possible to measure the following variables: date of egg laying, number of eggs, number of hatched nestlings, number of fledglings, and body condition of fledglings. Furthermore, blood of the nestlings was analyzed on day 14 after hatching.

Comparing the statistics from the uninfested nestboxes with those from the infested nestboxes gave the following main results: (1) a delay of egg laying by an average of 11 days in the infested nestboxes; (2) no reduction in clutch size in the infested nestboxes (in fact, there was even a slight, statistically insignificant excess of eggs in the infested nestboxes); (3) slight difference in hatching rates (94 percent in the uninfested nestboxes, 91 percent in the infested ones); (4) very clear difference in fledging rates (84 percent in the uninfested nestboxes, 53 percent in the infested ones); (5) significant degradation of the body condition of nestlings in infested nestboxes (as assessed by the ratio of body weight to tarsus length); (6) significantly lower hematocrit rate (ratio of volume of red blood cells to total blood volume) in nestlings of infested nestboxes; and (7) shorter dispersal distances after fledging by nestlings from parasitized nests.

This research shows that the presence of fleas in a nestbox clearly affects not only number of descendents of a pair of tits but also their quality. Quality of offspring is as important as quantity, because the subsequent survival of young birds is tightly linked to body condition. Among the life history features that result from natural selection in birds, timing of reproduction is one of the most important because it influences the probability that reproduction will succeed. If egg laying and hatching of young occur slightly too late, the fate of a clutch can differ greatly from that of a clutch laid at the optimal time because the food for the late clutch can be of inadequate quality or quantity. The delay in laying detected in infested nestboxes can therefore have important consequences. The various and cumulative effects of parasitism confirm that great tit reproductive success is sharply reduced if the parental pair nest in a flea-ridden nestbox.

The question that then arises is therefore, What steps can the tits take to avoid the negative impact of fleas? Richner, Oppliger, and Christe (1993) predicted that the choice of nestbox during the mating season can be influenced by the presence of fleas, if the birds were able to detect them. To test this prediction, they conducted the following experiment.

Pairs of nestboxes located less than a meter apart were offered to tits in three series (A, B, and C) of 15 pairs each (in the following description, F represents a nest or nestbox infested with parasites and U an uninfested nest or nestbox) (fig. 4.3). Although in nature, the birds themselves probably infest nests and nestboxes by carrying fleas, only the researchers infested these experimental nestboxes.

Series A: Each pair comprised a nestbox containing a nest U and an empty nestbox U (a nest consists of a collection of twigs destined to receive eggs, whereas the nestbox is the box containing the nest).
Series B: Each pair comprised a nestbox containing a nest U and an empty nestbox F.
Series C: Each pair comprised a nestbox containing a nest F and an empty nestbox U.

The experiment showed that the tits tend to select nestboxes containing previously constructed nests if both nestboxes lack fleas (series A), but when one of the nestboxes is infested, they tend to choose the uninfested nestbox, whether it contains a nest (series B) or not (series C). Even if the choices are not 100 percent in the direction indicated, the tits are able more often than not to avoid the most flea-ridden nests. This a case of avoidance behavior. To a degree, the tits succeed in closing the encounter filter.

Predicting Parasites

A change in a demographic parameter can constitute a defense mechanism. For example, selection to reproduce at an earlier age can allow host individuals to transmit their genes before the virulence of the parasite comes into full play to diminish reproductive success. Exactly this sort of

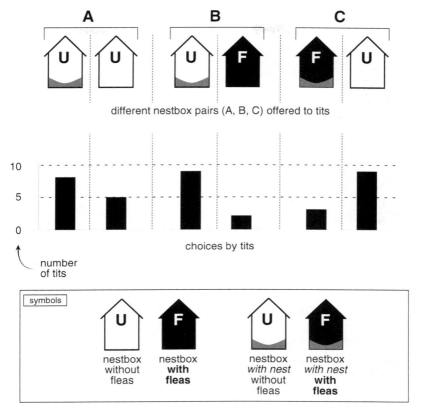

different nestbox pairs (A, B, C) offered to tits

choices by tits

number
of tits

symbols

nestbox
without
fleas

nestbox
**with
fleas**

nestbox
*with nest
without
fleas*

nestbox
with nest
**with
fleas**

Fig. 4.3. Experiment demonstrating the ability of tits to choose an uninfested nestbox (U) rather than one infested by fleas (F), even if the nestbox with fleas has a ready-made nest (and would therefore save some energy).

selection is seen in systems in which virulence is very pronounced. (For example, in a mollusc that runs the risk of being castrated by a trematode, it is better to reproduce while young than not to reproduce at all.)

We should recall at this point a fundamental trade-off that birds must confront independent of parasitism: this a question of whether to invest in either current or future reproduction. Clutch size is one life history trait that is the object of one of the most important trade-offs in the life of a bird. As noted by Rothstein (1990), clutch size determination is a classical topic in ecology and evolution. The compromise is between the size of the clutch and subsequent survival of the mother; is it better, in terms of successfully transmitting one's genes, to invest in a large clutch at the risk of being so exhausted by the effort of producing it that one does not survive to the next reproductive season, or to be less ambitious with respect to the size of the current clutch but to increase one's chances of surviving until the next spring? In other words, if the bird invests too much energy in current reproduction, it decreases the chances of being in good condition for future reproduction. Conversely, if it invests less in current reproduction, it maximizes its likelihood of substantial reproduction in the future. Optimal clutch size is the result of this trade-off.

If ectoparasites are present in the nest, they can interfere with this trade-off in the sense that the "right choice" between a large clutch now and many small clutches in the future can differ depending on whether there are parasites and, if there are parasites, which species these are.

Richner and Heeb (1995), on the basis of a series of studies of several species of birds and several species of parasites, concluded that optimal clutch size is not the same when different parasite species are present. If the parasite does not multiply during the course of nestling development (as is the case for fleas), there exists at the outset of the breeding season a certain stock of parasites, and this stock is shared among the nestlings. Therefore, the larger the clutch, the fewer parasites on average on each nestling (and thus the less its development is affected), as a result of a phenomenon of dilution. The "right choice" is then a large clutch. However, if the parasite multiplies during nestling development (as, for example, is the case for mites), the parasite stock grows proportionally with the number of nestlings and the advantage of a large clutch size disappears. A large clutch would incur the risk that the parents could not adequately feed a group of nestlings that is both too numerous and weakened by parasites. The "right choice" would then be a small clutch.

Of course, as Richner and Heeb (1995, 435) say, "whether females should adjust clutch size will largely depend on whether they can, when laying the clutch, predict the parasite load after hatching." For "prediction" by the females to be possible, parasite pressure would have to be relatively constant for several reproductive seasons, so that natural selection would favor the transmission of either "large clutch" or "small clutch" females, depending on the type and rate of parasitism. Richner and Heeb admit the possibility, on the part of the female, of ad hoc responses—that is, after the clutch is laid and in the face of effects of parasitism, some nestlings might be sacrificed or, on the contrary, some extra eggs might be laid. This would be a matter of behavioral correction of poor initial assessment of parasitism rate in determining optimal clutch size in the face of parasitism.

The Second Line of Defense

Compensating for Parasite Impact

When I speak of a second line of defense, readers likely think immediately of classic immunity: T lymphocytes, B lymphocytes, antibodies, and so forth.[3] Research on tits proves first of all that hosts can endeavor to compensate for parasite impact. Compensating for parasite impact constitutes a good part of the second line of defense, because this compensation occurs only if the infestation has occurred.

The presence of fleas entrains an entire series of consequences in the life of a nest. How are the nestlings, who are the main victims of the ectoparasites, going to react? How are the parents, confronted with an attack on their progeny, going to react?

Christe, Richner, and Oppliger (1996) asked three questions. First, Will parasitized chicks cry less because they are weakened (and therefore be fed less by their parents), or will parasitized chicks cry more (and lead their parents to make an extra effort to feed them)? Second, Will parents do everything in their power to compensate for the negative impact of the parasites, or will they sacrifice the current clutch to enhance their chances for future reproduction? Third, Is competition among the nestlings modified by the presence of parasites?

To answer these questions, Christe, Richner, and Oppliger made a detailed comparison between a series of parasitized great tit nests and an unparasitized series. Thirteen days after the eggs hatched, the researchers measured four variables: the fraction of time chicks spent calling each hour, the feeding activity of the parents, the weight of the chicks, and the weight of the parents. The results were as follows (fig. 4.4):

- Total time spent calling each hour was 18 minutes in uninfested nests and 39 minutes in parasitized nests; the answer to the first question is that parasitized chicks call more.
- The number of feeding trips by the father was 15 per hour in uninfested nests and 24 per hour in infested ones, whereas the number of feeding trips per hour by the mother did not differ significantly between uninfested nests (18) and infested ones (20); the answer to the second question is that one parent attempts to compensate for the impact of the parasite.
- Chicks in uninfested nests are fed more equitably than those in parasitized nests; the answer to the third question is that competition increases in parasitized nests.

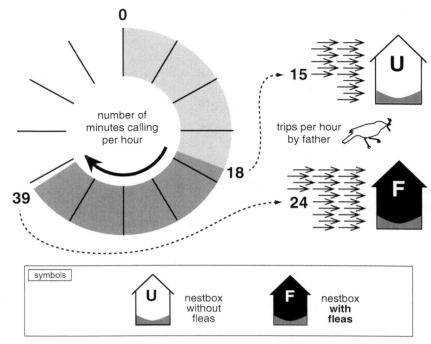

Fig. 4.4. Increase in calling time by parasitized nestlings (circle at left represents 1 hour) and increase in frequency of feeding trips by their parents (arrows at right each represent one trip per hour). (Modified from data of Christe, Richner, and Oppliger 1996)

Moreover, Heeb et al. (1998) showed a positive correlation between the total mass of the nest and the number of fleas (adults and larvae) found within it. There are several possible explanations for this correlation. For example, one can imagine that there is random variation in mass among nests and that tits are better at eliminating ectoparasites in nests in which the weight of the twigs and other nest material is low. However, Heeb et al. believe the best explanation is that increase in nest mass is an adaptive response by the tits; they compensate for the costs of parasitism, and especially the weakening of the nestlings, by producing better thermoregulation for them by constructing a heavier nest. In general, good insulation is important for successful fledging of a clutch, and this factor can become even more crucial when a clutch is heavily flea ridden.

Who Pays?

Tripet and Richner (1997b), comparing the cost of parasitism for blue tit (*Parus caeruleus*) nests infested and uninfested by *Ceratophyllus gallinae*, noted there was no apparent cost of parasitism for the chicks; in particular, body mass and survival rates of chicks were similar in infested and uninfested nests. Neither did they find evidence that the increased feeding activity in infested nests caused the parents to lose weight. At this stage of the research, one could have the impression that there was no cost of parasitism for either nestlings or parents, which seems unlikely. Normally, because fleas are responsible for an extra energetic expenditure by the birds, this cost would have to be paid in some way.

Richner and Tripet (1999) showed that this prediction is correct: parasitism reduces both the probability that the pair returns the next year and the number of young reared the next spring. So what is affected are the probability of survival and what is called the residual reproductive value. In figure 4.5, reproductive success in the second year is considered to be the product of the probability of return and the number of young reared in that year.

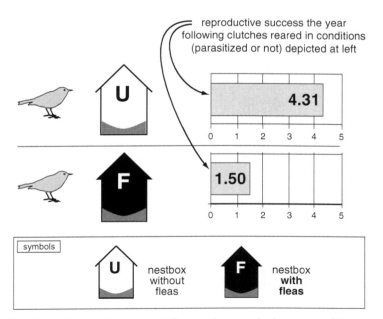

Fig. 4.5. How raising a clutch in year *n* influences the reproductive success of tits in year *n* + 1. Confidence intervals have been omitted. (Data from Richner and Tripet 1999.)

In the blue tit, therefore, it is the parents rather than their offspring who pay the cost of parasitism.

Reproduction and Immunity

Blue tit parents, in response to parasitism by fleas, do not merely give more food to their chicks, as shown in the experiments just described. It seems they can also transmit "immunological weapons" to them, as the following experiment suggests.

Heeb et al. (1998) selected a series of 46 nests and completely disinfected them (by microwaving them) at the beginning of egg laying. No living parasites remained. Next, half of those nests were left as is while the others were each infested with 40 adult fleas so that birds in the latter nests had contact with the parasite. Then, when laying ended, all nests were again disinfected. Finally, all nests were infested with 40 adult fleas. Thus, all 46 pairs of tits raised their young in infested nests, but 23 of them (group A) had contact with the parasite during egg laying, that is, about 10 days before the 23 others (group B), who contacted parasites only at the beginning of incubation.

All the results indicate that the birds in group A had better defensive abilities: the proportion of young surviving to fledge was significantly greater in group A than group B; the interval between hatching and fledging was shorter in group A; 16-day-old nestlings were heavier in group A; and the number of birds captured the following nesting season as 1-year-olds was greater in group A.

The explanation of the better tolerance by birds of group A of the same degree of parasitism is not known with certainty. However, the most reasonable hypothesis is that group A females had produced antibodies and transferred them to their eggs, thereby allowing their nestlings to mount a more effective immune response against the fleas.

Immunity and Its Compromises

The Reallocation Principle

The establishment of an immune system line of defense is expensive for the tits, as it is for any species. When a host individual is infected by a parasite, that parasite takes a part (small or large, depending on the species) of the host's resources. A pathogenic effect ensues—for example, weight loss, shorter movements, and so forth. This pathogenic effect, because it lowers the reproductive success of the host (that is, is virulent), selects for defense mechanisms in the host species.

These mechanisms have costs, as does any other function of an organism. The theory of life history traits (Stearns 1988) posits that any energy allocated to one function is no longer available for any other function (I have already discussed life history traits in chap. 3 with respect to parasites rather than hosts). It is as if an individual has an "energy budget" and is constrained to balance it, just as a city or nation is.

Mobilizing an immune response imposes a cost, then, which is deducted from the total sum of resources available to the individual, and that part of the budget devoted to this defensive response could have been used for other functions if the individual had not been infected. When an individual is parasitized, there is thus a resource reallocation. We can expect that this reallocation is such as to minimize the impact of the parasite on host reproductive success. Research by Richner's group gives examples of resource reallocation. This research shows there are compromises between the quality of the immune response and both number of eggs laid and number of nestlings.

Clutch Size

In the vicinity of Lausanne is found an endemic endoparasite of tits belonging to the genus *Plasmodium* (related to the *Plasmodium* that causes human malaria). This is a unicellular, mosquito-transmitted parasite that penetrates the red blood cells (red corpuscles) of birds. The growth of the parasite ultimately bursts the red blood cells. The parasite weakens birds but is so far not known to kill them.

Oppliger, Christe, and Richner (1996) showed there is an interaction between clutch size and resistance to the *Plasmodium*. They removed the first two eggs laid in each of 35 nests of the great tit (*Parus major*), while the clutches of 41 nests were left intact to serve as controls. The females of the group from which eggs were removed compensated for the loss by laying, on average, one extra egg. In each of the 35 experimental nests, the female therefore had one fewer nestling to rear, on average, but she had laid one more egg than the control females. Fourteen days after a brood's hatching, blood was taken from the adult female to determine the prevalence of avian malaria. The results (fig. 4.6) show a highly significant increase in rate of malaria in adult females (50 percent in the experimental group vs. 20 percent in the controls) and no difference between the adult males of the two groups. We conclude that laying even a single extra egg suffices to render the females more vulnerable to some parasites. This experiment furnishes one more argument for the proposition that clutch size has a cost and that paying this cost lowers the quality of defenses and influences survival of a bird, and therefore its future reproduction.

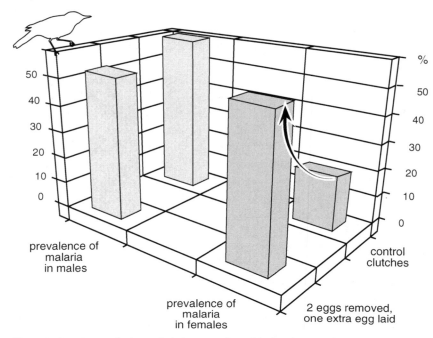

Fig. 4.6. Prevalence of avian malaria increases (*arrow*) in female great tits when the egg-laying effort increases.

Number of Nestlings

Richner and his colleagues further tested whether allocation of resources to parental care also affected the prevalence of this disease (Richner, Christe, and Oppliger 1995). To this end, they manipulated number of nestlings so as to increase artificially the energy the birds devoted to feeding their young. They obtained two groups of nestlings: (1) reduced groups, from which two nestlings were removed the day they hatched; and (2) enhanced groups, to which two nestlings from other clutches were added the day of hatching.

The researchers then evaluated feeding rate, which turned out to differ between the two groups. Although, strangely, females showed the same rate of visiting their nestlings in the reduced and enhanced groups, the males, by contrast, increased their number of feeding trips by 50 percent in the enhanced groups relative to the reduced groups. Now, there is a disturbing correlation between the increased activity of males and the prevalence of *Plasmodium*. Although the parasite is present in 38 percent of the males in the reduced group, it is found in 76 percent of the males in the enhanced group (fig. 4.7). The most likely explanation is that allocating resources to intensive feeding diminished allocation to immunity such that the over-worked males were more susceptible to the parasite transmitted by mosquito bites. However, another possible explanation is that because they made more feeding trips, the males in the enhanced group were exposed to mosquitoes for a longer time.

A related question is, Why do the males increase feeding rate, but the females do not? Applying the principle of compromise among different resource allocations, I propose two answers.

I have explained how a bird is presented with the choice between investing heavily in the current reproductive period or conserving its forces for future reproduction. The difference in feeding effort between females and males suggests that female tits are more likely than males to invest in future reproduction. By not increasing their energy expenditure for enhanced clutches, they are, in a sense, betting that this response will pay off in the end in increased transmission of their genes. Because natural selection has produced this behavior, it is probably because the

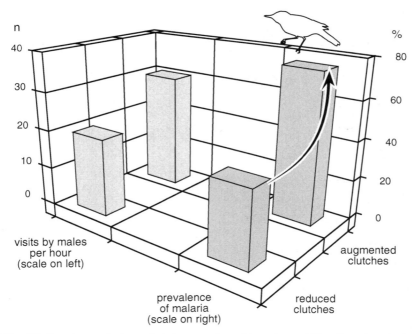

Fig. 4.7. Prevalence of avian malaria increases (*arrow*) in male great tits with an increase in their parental activities.

probability of future reproduction (the next egg-laying season) is increased for females who behave this way. The fact that males give their all for the current clutch signifies, on the contrary, that their chances for future reproduction are probably not great. It is better in these circumstances for them to sacrifice everything in the attempt to transmit their genes today and risk not transmitting them later. This male tit behavior confirms the old proverb, A bird in the hand is worth two in the bush.

Christe, Richner, and Oppliger (1996) suggest a different answer to the question. In tits, females normally divorce males if a clutch has little chance of successful fledging. The staggering investment of males in feeding the enhanced clutch might be a means of "showing good faith" and avoiding a divorce at all costs.

Resistance to Insecticide versus Resistance to Bacteria

The principle of resource allocation applies to more than just resistance to parasites and investment in reproduction. Berticat et al. (2002) studied bacterial infection of mosquitoes subjected (or not) to an insecticide. Mosquitoes of the genus *Culex,* as do many insects, harbor bacteria of the genus *Wolbachia* (discussed in chap. 3). The use of organophosphate insecticides entrains, in mosquito populations, selection for genes causing increased production of certain enzymes (esterases). Research has shown that mosquito populations that have evolved resistance to insecticides are much more heavily infected by *Wolbachia* than are susceptible populations. Here the compromise is between investment in resistance to bacteria and investment in resistance to the insecticide.

Missing Parasites

That parasites play a role in regulating host populations is not questioned today. This realization has led to the suspicion that a diminution of parasite pressure can lead to a "deregulation," allowing some introduced species to become invasive (cf. Torchin et al. 2003).

Any population is subjected to pressures from populations of other species with which it shares a location. If an introduced species finds fewer competitors or predators in its new home, its reproductive potential can be enhanced and its population can quickly proliferate. Similarly, if an introduced species has fewer parasites in its new home than in its native range (the enemy release hypothesis), it is likely to have greater reproductive success (Mitchell and Power 2003; Torchin et al. 2003).

An introduced species can have fewer parasites in its new range than in its native range for two reasons. First, because parasites often have complex life cycles that require several successive host species, it is possible that some of these necessary hosts are absent from the new range. Second, when a species is introduced, it is generally as a very small fraction of the original population. Geneticists speak of a "founder effect," whereby only a part of the genetic diversity of the original population arrives with the founders of a population in a new range. Now, the same sort of founder effect may hold for infection by parasites: there is no reason to think that a sample of host individuals will have all the parasites of the entire host population.

Controversy: The Profession of Human Host

In what species has the greatest number of parasite species has been detected? Answer: *Homo sapiens*. Humans garner the dubious honor of exercising the profession of host in the most accomplished manner. Is there an explanation for this remarkable but unwanted performance? In fact, there are two.

The first is that humans have been studied for parasites (both macro- and microparasites) more than any other species, for obvious reasons related to health.

The second is that, unfortunately, it seems that, aside from this sampling bias, *Homo sapiens* is truly host to an incredibly elevated number of parasites (relative to other species), whether one counts only parasites in the strict sense (protists and metazoans) or adds in viruses, bacteria, and fungi. This richness is due to the quantitative importance of lateral transfer—that is, recent acquisition of parasites that evolved in animal lineages other than those that led to modern humans.[4]

In the parasite fauna of any host, we can delineate two categories: (1) parasites that evolved in the lineage of that host and consequently were inherited by that host from that host's ancestors; and (2) parasites that evolved in different lineages and arrived in the current host by lateral transfer.[5]

It appears that in most cases, parasites of the "inherited" type are more numerous than recent transferees. These latter can constitute a substantial number of species only if the host species has recently undergone changes in its ecology and/or behavior. Or if its ecosystem has been invaded by "new" species carrying "new" parasites.

Matters are quite different for the human species. It appears that hominids have been a godsend for many parasites by offering them many transfer opportunities. We must recall that the recent evolution of primate ancestors of modern humans was marked by profound changes in ecology and behavior.

With respect to ecology, the change from an arboreal to a terrestrial existence led to encounters with "new" parasites for several reasons:

- Hominids were exposed to new biting parasite vectors such as ticks and some species of mosquitoes and horseflies, as well as to infestation by parasites of vertebrates living at ground level.
- The change in lifestyle was probably accompanied by contact with an ever greater number of aquatic habitats and especially bodies of stagnant water where parasites accumulate.
- Our ancestors illustrated the rule according to which species living on the ground have a more markedly exploratory temperament (therefore running more risks) than arboreal species, and they have undertaken ever more distant migrations and moved into an ever-increasing number of new environments.
- Hominids, leading to *Homo sapiens,* have consequently invaded practically every ecosystem on the planet.

As for behavior, for reasons of hygiene, fishing, or agriculture, humans have continually increased their contact with water, and the invasion of new ecosystems has been accompanied by new foods. When we consider that many parasites have tight relationships with water and that others use food chains and food webs for their transmission, we understand to what degree each new use of an aquatic habitat or each new food can constitute a potential danger to health.

Changes in ecology and behavior have led to encountering many parasites that had not previously parasitized primates. Among all the infective stages, some fraction were unable to develop in humans, either because the parasites were not adapted to exploiting the new arrival or because the new arrival had immune resources that destroyed them. But the others turned out to be able to "get to humans." Possibly some got to humans right away, the first time they encountered them, whereas others did not succeed in this new host until after a bout of natural selection imposed by the human immune system.

Figure 4.8 shows three processes that could have led humans to capture parasites from other host species. Humans became hosts of *Clonorchis* by eating poorly cooked fish because the "natural" definitive hosts of these trematodes were fish-eating mammals. Humans became hosts of *Schisto-soma* by bathing in water infested by their cercariae, which originally penetrated other mammals. They became hosts of *Onchocerca* by being bitten by blackflies (Simuliidae) that had previously transmitted their parasites to bovids.

The chosen examples concern relatively ancient transfers. (The passages probably occurred one or two million years ago, which is not a long time on an evolutionary time scale but is very long compared to the life of an

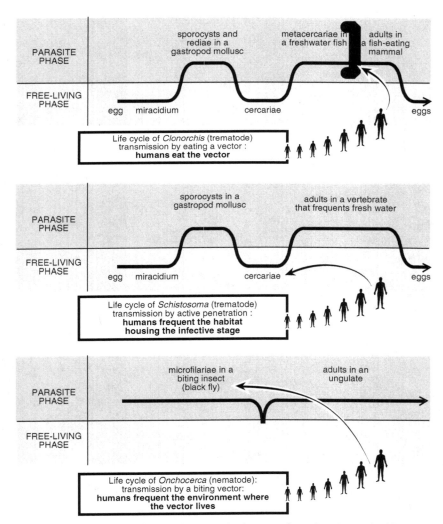

Fig. 4.8. Three examples of probable capture by humans of parasites that evolved in non-primates.

individual human.) The last 20 years of the twentieth century have shown that such transfers can still be produced today, and even more and more frequently because of ever more frequent human behavioral changes. The case of the HIV virus of AIDS (see chaps. 3 and 10), very likely passed from chimpanzees to humans, and that of mad cow disease (see chap. 10), passed from sheep to cows by means of feed constituted partly of meat and thence probably passed to humans, testify to the currency of this process.[6]

A very unexpected transfer has recently been revealed. We knew that the tapeworm *Taenia saginata* is transmitted to humans who eat poorly cooked beef, while *T. solium* is transmitted by the consumption of poorly cooked pork. It therefore seemed likely that our ancestors acquired these parasites when they domesticated cattle and pigs and began eating beef and pork. In fact, this is not at all how these transfers happened. Research on the *Taenia* genome (Hoberg et al. 2000, 2001) shows that the ancestor of *T. saginata* is found in felids (cats) and that of *T. solium* in hyaenids. All hyaenids and many cats are African. Felids and hyaenids became infested with *Taenia* by eating ungulates on the savanna. The parasites were first passed from these wild carnivores to humans. Later, humans contaminated domesticated cattle and pigs, which then became the vectors of these parasites. The latter did not go from domesticated animals to humans but from humans to domesticated animals.

When a parasite passes to a new host, the result can be a simple addition of a new host to the one or several the parasite exploits already. In other cases, speciation can occur so that a new parasite species becomes differentiated on the new host. These alternative processes are both represented for parasites passed to humans in figure 4.8:

- There was no speciation in *Clonorchis; C. sinensis* infests many fish-eating mammals as well as humans.
- In schistosomes, one species, *Schistosoma haematobium*, has differentiated from ancestral parasites of bovids; another, *S. mansoni*, has differentiated from ancestral rodent parasites.[7] In the Far East, by contrast, *S. japonicum* resembles *Clonorchis sinensis* in that it indiscriminately exploits humans and many other mammals that frequent aquatic environments.
- In *Onchocerca*, the human parasite *O. volvulus* is descended from a species attacking livestock, but nowadays it parasitizes only humans.[8]

To assume the profession of host has surely been one of the main features of the survival of small bands of primates that resembled humans hundreds of thousands of years ago. Of course the risk of encountering large predators also weighed heavily on our ancestors, but parasites must have played an important role, indeed the major role. One can imagine that parasitic diseases, adding up to strike hominids (above all when the latter conquered new habitats), have long succeeded in stemming the growth of human populations. It is also possible, conversely, that the spread of humans outside the tropics was aided by the fact that, in the temperate zones, the burden of parasitism was much lower. Even the glaciations acted against certain parasitic diseases, while the invention of fire and its use in cooking food killed many infective stages.

I reserve for chapter 9 a discussion of the way in which scientific progress has made humans into the most professional of all host species in the fight against infectious and parasitic diseases.

The Profession of Mutualist

The bringing together of genetic information from two unrelated sources can certainly generate novelty.

◊

JOHN MAYNARD SMITH
(1989)

How Parasites Are Exploited

The Strategy of Subensembles

When individuals of two species whose evolution has previously been independent associate with one another so that each benefits, they are said to engage in mutualism or symbiosis.[1] When this type of association is mentioned, the example that often comes to mind is that of the hermit crab and its anemone.[2] This small marine crustacean, which protects its soft abdomen in an old gastropod mollusc shell, carries around one or more sea anemones permanently affixed to that shell. Each species benefits from this association: the hermit crab gains protection from predators, while the anemone gains a degree of mobility. This sort of association is common.

In this example, mutualism seems to be just a picturesque, even anecdotal, aspect of biological evolution. In fact, mutualism has played and continues to play a major role in the biosphere.

Arthur Koestler wrote what is called the parable of the two watchmakers. Each builds a watch, but they proceed differently. The first, named Bios, begins by making units of ten pieces. Next he assembles these units into groups of ten and continues in this fashion until the watch is finished. The second watchmaker, named Mekhos, builds watches piece by piece, with no intermediate units. From time to time, a disturbance scatters the pieces being assembled, so he must begin anew.

We quickly recognize that Bios's strategy is better than that of Mekhos. Bios loses at most the effort that went into assembling one of the ten-piece units, whereas Mekhos risks having to begin the entire assembly over again from all its components. Egbert Leigh and Thelma Rowell (1995), who describe this parable, argue that, if disturbances are sufficiently frequent, only Bios has a chance to complete his watch. Transposing this story of watchmakers to the history of life, these authors conclude that the most spectacular increases in complexity in the history of life result from the association of preexisting subensembles; this association produces a higher level of complexity. Maynard Smith and Szathmary (1997) concur when they develop the concept of *major transitions* at certain moments of evolution. In biology, a subensemble is a living being whose association with another subensemble (another living being) leads to a new entity, which constitutes a third living being. In this conception, the strategy of subensembles, as I call it here, would therefore be as fundamental for biological evolution as the process of natural selection. In the controversy I discuss at the end of this chapter, I show that subensembles do not always live in perfect harmony.

Commensalism and Phoresy: Transitional Situations

There exist associations that, although obligatory for one of the partners, *seem* to be completely neutral for the other, neither costing it anything nor benefiting it. I first explain what commensalism is by one example chosen from many of this type.

In the Gulf of Mexico lives a sponge, *Stellata grubii,* whose folds shelter a small oyster, *Ostrea permollis.* The sponge lives on sandy bottoms where solid substrates are rare. It is spherical and reaches 30 centimeters in diameter. The oyster is approximately 3 to 4 × 2 centimeters. It has the classical morphological and anatomical features of an oyster, but the light yellow valves of the shell seem relatively fragile. The main exchanges between the oyster and its environment are depicted in figure 5.1.

Research on this association (Forbes 1966) shows that (1) some *S. grubii* have no oyster symbionts; (2) no *O. permollis* are attached to substrates other than the sponge; (3) a sponge that does not harbor oysters seems to have no advantage or disadvantage relative to one that does; and (4) the oysters live by filtering plankton and never eat sponge tissues.

The characteristics of this association suggest, on the one hand, that the oyster cannot live without the sponge whereas the sponge can live without the oyster, and on the other hand that the oyster does not damage its host in any way. Predation tests demonstrate the benefit the oyster draws from

Fig. 5.1. Relationships among the oyster *Ostrea permollis*, the sponge *Stellata grubii*, and the surrounding seawater. The entering spermatozoids are those coming from other oysters, whereas the departing spermatozoids are those produced by the pictured oyster.

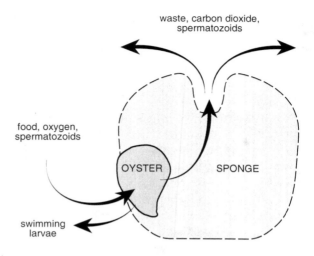

this association. For example, gastropods of the genus *Murex*, which pierce the shells of bivalves to feed on their innards, cannot attack *O. permollis* embedded in sponge tissue. The sponge thus offers some protection against predators, but it is not total. Associations of this type are very numerous. Just among bivalves, some species are associated with corals, others with worms, and so forth.

These examples show that what is typically called commensalism is only a subset of either parasitism or mutualism in which the cost or benefit for one partner (the sponge in this case) is probably so minute that it cannot be measured. However, it seems logical to suppose that the burden of the oysters, which are often massed on sponges, causes the sponges some loss of reproductive success, even if only by limiting the sponge surface exposed to the surrounding sea. Neither is it impossible that the oysters confer a benefit on the sponge, limited though it may be. For instance, it has been suggested that the water current created by the oysters aids sponge nutrition—that is, that the current generates the opposite effect to that just posited. As we see, classifying this association as commensalism results more from the difficulty of measuring the costs and benefits than from a true demonstration that the association is neutral for one of the partners.

A similar difficulty in distinguishing costs and benefits has led to many associations being classified as commensalism. Describing the association between a pilot fish (*Remora*) and a shark, Maurice Caullery (1950) wrote that the first follows the second just as a piece of iron follows a magnet, instantly obeying the incessant and irregular changes of route by the shark. This author also refers to the fishes and shrimp that live amid sea anemone tentacles, the amphipods (crustaceans) sheltering in the bell of a jellyfish, and many similar relationships. In each instance, the benefit seems evident for one partner (most often, it is a matter of protection), while the cost or benefit for the other partner seems weak if not nil.

Consider now a second transitional stage toward mutualism, phoresy. Françoise Athias-Binche has devoted much research to phoretic mites. What are these?

Mites are arthropods, generally small, whose adult stage has four pairs of legs, as do spiders. Some, like the oribatids, are free-living, whereas others, such as ticks, are parasitic. Still others punctuate their free-living life with periods of phoresy. That is, they cause themselves to be transported by individuals of another species, in principle without harming that individual. The reason for this transport is as follows: many mites live in specialized and fragmented microhabitats, patches of which are often separated by distances that simply cannot be crossed by species that are not very mobile. Such habitats include hollows in tree trunks, decomposing fruits or mushrooms, and cowpats. If the mites attach to other animals (insects, for example) that visit these microhabitats but are able, thanks to their wings, to traverse long distances, the problem of dispersal is resolved for the mites. Mites are not the only phoretic species, but they are certainly the best studied.

I will cite some examples. Species of *Allodinychus* live in dead wood, a resource shared with many beetle larvae. In the spring, these larvae transform to adults, which emerge from the wood, then disperse to search for suitable sites in which to lay their eggs. It is then that the mites attach themselves to the feet of the beetles and are themselves dispersed. They remain motionless during the flight and detach spontaneously when a beetle lands on a new piece of dead wood.

Species of *Naiadacarus* live in small pools of water that accumulate in hollows of some tree species. This is the habitat in which fly species in the genus *Eristales* lay their eggs, which hatch to become aquatic larvae. When the flies contact the water in order to lay their eggs, the *Naiadacarus* grasp their legs and are then transported in the same way as *Allodinychus*. When the flies find another water-filled tree hole, the mites drop off.

Phoresy poses a problem similar to that of commensalism in that costs and benefits must be measured. Athias-Binche and Morand (1993) observe that the total inactivity during the transport phase distinguishes phoresy from parasitism. They note, however, that on the one hand no benefit accrues to the transporting host, while on the other hand the mites, even though they do not feed on their host, must have a negative impact, weak though it may be. The mites constitute a burden for the insect, and they probably make it less aerodynamically efficient. Athias-Binche and Morand report specific instances in which an abundance of phoretic passengers can render flight difficult for an insect. They cite the case of a fly 1.85 millimeters long transporting 18 mites each 0.31 millimeter long, which represented 13–14 percent of the volume of the fly. It therefore appears justifiable to consider phoresy as practiced by *Allodinychus* and *Naiadacarus* as a form of parasitism imposing minuscule costs.

Athias-Binche (1990, 1994) showed that, beginning with this "typical" phoresy, there are two possible evolutionary paths. One is toward parasitism pure and simple (the passenger feeds itself at the expense of the carrier), the other toward mutualism (the passenger somehow pays the carrier). The services the mites render to the insects that carry them are of several sorts: they can clean their nests, rid them of parasites, attack their enemies. An example of evolution toward mutualism is the case of *Parapygmephorus,* which is transported by bees. The mites detach from their host when the host constructs its brood cells (in which the bee larvae are reared) and lay their eggs beside the bee larvae. The mites are very busy housekeepers within the brood cells, consuming detritus of all sorts, which hinders fungal growth.

One can ask why the phoretic mites that tend toward mutualism bother to "pay" for their transportation. As in the rest of the biotic world, this is not a form of freely given generosity. The mites draw from their activity "on behalf" of the host either a direct benefit (for example, cleaning the nest and consuming the detritus may augment their own reproductive success) or an indirect advantage (for instance, their activity may increase the probability of survival for the host descendants). Whichever, their activity can result only from natural selection.

Examples of Mutualism

The Mercenary Viruses

We usually have a very negative view of viruses. We think of the word *virulence;* we think of the AIDS virus, that of smallpox, whooping cough, measles, hepatitis, shingles. Moreover, practically all animals and plants, not only humans, can be victims of viruses that are often very pathogenic. Insects figure among the potential victims. The proof of this fact is that viruses are in the first rank of tools used in the biological control of plant-eating insects that attack agriculture.

Some insects, however, are infected by viruses that are true mutualists because they are absolutely essential to their hosts. Implicitly admitting that these mutualists almost surely derive from what had formerly been pathogens, experts on these associations use the elegant and evocative term *domesticated viruses.*

The "domesticators" of these viruses are wasp parasitoids that lay their eggs in the bodies of larvae of other insects, for example, butterfly caterpillars.[3] We can guess that the problem these parasitoids must solve is to overcome the defenses of hosts that are, in principle, perfectly able to encapsulate them and thus to kill them. This encapsulation defense of insects relies on the activation of what is known as the phenoloxidase system. Phenoloxidases are enzymes present in animals in the form of prophenoloxidases, for which a chain of activation depends on many other enzymes. The first reaction in the chain is triggered by the intrusion of "non-self." The final reaction produces several compounds, the most important of which is melanin, which is deposited around the intruding entity and therefore is part of the encapsulation. Any interruption of the chain causes the entire resistance mechanism to halt.

To attain this cessation, the parasitoid sometimes injects immunosuppressive substances along with the egg. In other cases, it uses viruses, which rush to the aid of the parasitoid's offspring. That is, viral particles are injected along with the egg. The best-known viruses of this sort are called polyDNA viruses. They are integrated into the genome of the parasitoid for the majority of the latter's life. When the wasp reproduces, the viral genomes again become autonomous in the wasp ovaries, where they produce viral particles that are injected into the host with the genital fluid that surrounds the wasp eggs. These viral particles do not multiply in the host caterpillar, but they completely neutralize its defenses.[4]

Mutualistic Bacteria of Entomophilous Nematodes

Species of *Xenorhabdus* and *Photorhabdus* are pathogenic bacteria of insects. A vector injects them into the hemocoel of the insect (just as the *Plasmodium* of malaria is injected into the blood of humans by mosquitoes). The vectors of these bacteria are soil-dwelling nematodes, in the genus *Steinernema* for *Xenorhabdus* and the genus *Heterorhabditis* for *Photorhabdus*. These bacteria are not very choosy and attack many host species; they are highly adapted to insects and are completely innocuous for vertebrates. When they have penetrated an insect, they multiply by millions and kill their host in two to three days. They are pathogenic only if they reach the hemocoel. The digestive tract kills them if they are ingested, so they are not dangerous if they enter the host by this route.[5] *Photorhabdus* owes its name to the fact that species in this genus emit light both within insects and in laboratory culture.[6]

Although the pathogenic agents are very different, and so are the hosts, it is instructive to compare the transmission of *Plasmodium* to that of *Xenorhabdus* and *Photorhabdus*. In fact, aside from some striking differences in details, the overall systems are interestingly similar (fig. 5.2). The mosquito that transmits *Plasmodium* injects it through the skin of a human in the course of a blood meal. Having reached the blood, the parasites multiply. Then the parasites are transmitted to another host by another mosquito and the cycle begins anew.[7] The nematode that transmits *Xenorhabdus* or *Photorhabdus* injects it into the insect by perforating the insect cuticle (or it liberates them internally in the insect by passing to the digestive tract). Up to this point, one could simply superimpose this life cycle on that of *Plasmodium*, but the first important difference is that the nematode vector injects itself along with the bacteria. In fact, it injects itself first and does not release the bacteria until later. The evolution of these bacteria is highly original because it is narrowly linked to that of the nematodes, to the point that we speak of a "nematode–bacterial" complex. I will first examine the biology of the nematode.

It is the larval stage L3 (that is, the third stage after the egg, with each stage separated from its predecessor by a molt) that penetrates the insect. Then a new molt leads to stage L4. Yet another molt and the nematode is an adult, either male or female. After copulation, the females lay their eggs in the hemocoel of their host, and the proliferation that results from one or two successive generations leads to an enormous infrapopulation of 150,000 to 250,000 individuals in each insect.[8] When no more resources are available in the insect, the parasite breaks out in the form of L3 larvae that therefore find themselves in the soil and seek new hosts.

The cycle I have just described would not be very original if the bacteria injected by the nematode did not play an essential role (Boemare et al. 1997; Brillard et al. 2001). Note first that, at the beginning, the bacteria are found in the nematode digestive tract but cause the nematode no visible harm because they do not feed and do not multiply. Even better, in certain nematode species, the bacteria are sheltered in a specialized intestinal vesicle, a fact that already suggests more of a mutualistic relationship than a parasitic one. Noël Boemare speaks of a bacterial "sanctuary."

From the moment the nematode penetrates its insect host, the nematode and the bacteria establish a tight collaboration to exploit what might be called their common host.

The nematode at first allows the activity of the antibacterial peptides secreted by the insect in response to the aggression it has undergone.[9] These peptides "cleanse" the insect of the microorganisms that the nematode

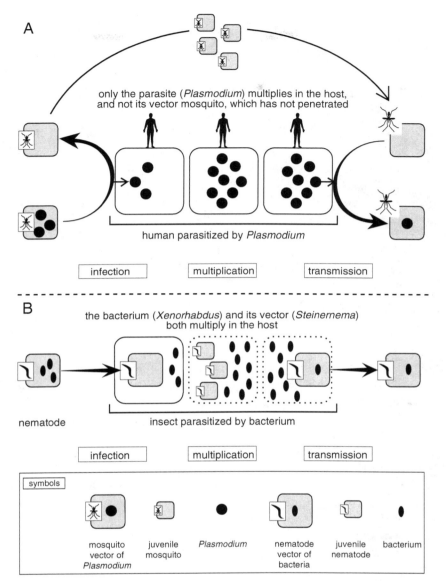

A

only the parasite (*Plasmodium*) multiplies in the host,
and not its vector mosquito, which has not penetrated

human parasitized by *Plasmodium*

| infection | multiplication | transmission |

B

the bacterium (*Xenorhabdus*) and its vector (*Steinernema*)
both multiply in the host

nematode   insect parasitized by bacterium

| infection | multiplication | transmission |

symbols

| mosquito vector of *Plasmodium* | juvenile mosquito | *Plasmodium* | nematode vector of bacteria | juvenile nematode | bacterium |

Fig. 5.2. Modes of transmission of a *Plasmodium* (A) and bacteria of the genera
Xenorhabdus and *Photorhabdus* (B). The *Plasmodium* life cycle is very simplified (sexuality is
not shown). The quantitative aspects are deliberately omitted from the two diagrams.

might have introduced accidentally while penetrating from the outside; then the nematode neutralizes the insect defenses by emitting substances that have an immunosuppressive effect and liberates the *Xenorhabdus* or *Photorhabdus* that it so far has sheltered in its digestive tract.

We see that the hemocoel of the insect host is successively disinfected and disarmed by the nematode, which thus paves the way for the development of the bacteria it is going to liberate. This whole process occurs as if the insect defenses do not recognize the nematode–bacterial complex.

The bacteria, after having been liberated by a nematode, complete the immunosuppressive action of the nematode by inhibiting the phagocytositic functions of the insect hemocytes;[10] they produce antibiotics that act against subsequent invasions of the insect by bacterial rivals; and they multiply profusely, using the resources available in the insect body.

The insect, as indicated earlier, dies two or three days after infection. Several days later, the nematodes, having greatly multiplied thanks to the resources furnished by the insect tissues and by the bacteria themselves,[11] leave the insect, which is by now reduced to just its cuticle. Before leaving the insect corpse, each nematode larva "recruits" about a hundred bacteria, which it carefully conserves in its digestive tract and which allow it to inoculate a new insect host.

The comparison between the *Plasmodium* of malaria and *Xenorhabdus/ Photorhabdus* allows us to highlight the characteristics of the relationship between a vector and the infectious agent it carries. The relationship between *Plasmodium* and its hosts, both humans and mosquitoes, is parasitic; it has never been demonstrated that the *Plasmodium* does anything good for the hosts. The relationship between the bacteria *Xenorhabdus* and *Photorhabdus* and the insect is also parasitic. By contrast, the relationship between the bacteria and the nematode is mutualistic: the bacteria take a bare minimum of resources from their vector,[12] collaborating with it to neutralize perfectly the insect defenses and acting as a teammate to exploit the insect. The fact that the nematode (vector) and bacteria multiply side by side in their victim is the most unusual trait of this association.

The notion of reciprocal benefit, which defines mutualism, is clearly illustrated: the benefits for the bacteria are the transmission from insect to insect and dispersal. The benefits for the nematode are the toxemia and septicemia that cause the insect to die and the predigestion of the insect organs. The costs appear to be negligible or nil. Even the fact that the nematodes consume a large part of the bacterial biomass should not be construed as a cost to the bacteria. The problem for the bacteria is that they need their genes transmitted, not that they need all individual bacteria transmitted.

Species of *Steinernema* and *Heterorhabditis* exist on all the continents. All *Steinernema* harbor *Xenorhabdus*. All *Heterorhabditis* harbor *Photorhabdus*. All indications are that the mutualism is ancient and has given rise to a process of cospeciation between nematodes and bacteria. Even more remarkably, it has been suggested that *Photorhabdus* may have had a marine origin, because the genes for their bioluminescence have high percentages of sequence homology (up to 85 percent) with those of luminescent marine bacteria. The question How did the ancestors of extant *Photorhabdus* get from the oceans to terrestrial ecosystems? seems not to have produced any cogent hypotheses yet.

I can add that the biology of *Steinernema* and *Heterorhabditis* resembles that of insect parasitoids in many ways. One main convergence is that they act like parasitoids, because the host is ultimately killed. Another is that they have recourse to a third partner in order to exploit their host. Like insect parasitoids, *Steinernema* and *Heterorhabditis* attack insect pests of agriculture, and they can therefore be exploited as biological control agents of these pests.

Pollinating Wasps

Plants are stationary. This trait is not very practical with respect to exchanging genes between individuals, unless the gametes themselves are mobile. The mobile gametes of plants are pollen grains, bearers of the haploid male nucleus. They achieve mobility in various ways.

Transport by water is exceptional, but aerial transport is common. In springtime, puffs of yellow powder escape from conifer forests; these are clouds of pollen. However, wind transport (anemophily) is a wasteful process; it is probably necessary to produce billions of pollen grains in order to ensure that one of them lands on the stigma of a female flower, produces a pollen tube, assures fertilization of an ovule, and therefore produces a seed.

Because plants are stationary and surrounded by mobile animals, there is an excellent complementarity of abilities that evolution has not failed to exploit. From this complementarity arose pollen transport by pollinators: birds, bats, and especially insects. All evidence points to the fact that the diversification of flowering plants on the one hand and pollinating insects on the other, which occurred for both groups about 90 million years ago (Machado et al. 2001), were linked (see the controversy discussed at the end of this chapter). Pollination by animals is a mutualistic process in the sense that there is an exchange of resources: the insect offers the plant its mobility, while the plant pays for this resource either with nectar or with part of the pollen itself.

Most of these mutualistic associations do not qualify as persistent interactions, however. In fact, when a honeybee or bumblebee flies from flower to flower seeking nectar, the plant–insect interaction lasts at most several seconds; this situation is not very different from that of a swallow catching fly after fly in the course of its flight, which is an act of predation. However, there are several types of plants whose pollination is also performed by insects, but in which the "predation" of pollen is replaced by a truly parasitic episode: the insect acts as a parasite of the reproductive apparatus of the plant and carries its pollen in exchange. Dufay and Anstett (2003) list a total of thirteen such associations. The best-known plants offering part of their seeds in exchange for pollination are yuccas, the alpine plant *Trollius europaeus,* and figs. I will discuss figs in more detail. Their pollinators are small wasps (hymenopterans) of the family Agaonidae.

I will try to recount the history of the relationship between figs and agaonids as if we were present during its evolution. Perhaps part of this reconstruction is arbitrary. At the outset I state that this history is a succession of hypotheses, of course, but of strongly supported hypotheses. The following paragraphs are largely inspired by the work of Finn Kjellberg and his colleagues (for example, see Kjellberg et al. 1987; Anstett, Hossaert-McKey, and Kjellberg 1997).

Toward the end of the Mesozoic era (approximately 65 million years ago), the ancestors of figs arose. Surprise—they were fig trees without figs, at least the figs we know today. They had inflorescences—that is, collections of flowers—but these were highly visible from the outside. This was the first stage in a curious evolutionary pathway (fig. 5.3). Millions of years passed, and the stalk that bore the flowers flattened, becoming a rounded receptacle, a bit like that of an oxeye daisy. This was the second stage.

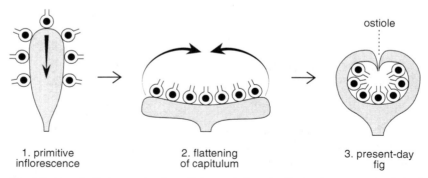

| 1. primitive inflorescence | 2. flattening of capitulum | 3. present-day fig |

Fig. 5.3. Hypothetical reconstruction of changes in form leading to the present-day fruit of figs. Male flowers are not depicted.

Finally, the receptacle closed in upon itself like a cognac glass with an opening so constricted as to be but a point.[13] This is the third and last stage. The constricted opening forms the ostiole. The flowers (there are separate male and female flowers) are imprisoned in the interior of what we now call a fig; there can be up to 7,000 flowers in a single fig.

And the agaonids? These tiny wasps initiated their relationship with the fig ancestors very long ago, and at the beginning the relationship was probably parasitic. Thanks to their long ovipositor, the wasps were able to lay their eggs inside the plant, thus assuring food and shelter for their offspring. Agaonids, like all parasites, were specialized, in this case for the female flowers of the ancestral figs. After several weeks or months, a young wasp emerged from a female flower in which an agaonid had laid an egg, and the cycle began anew. Now let us jump forward in time and look at the world of contemporary figs.

There are around 700 fig species, mostly in the tropics. Each species is associated with one species of agaonid wasp, with very few exceptions: there are 700 fig–wasp "pairs." At first glance, the agaonids appear to be ordinary parasites. However, if we look more closely, we see they all have the same peculiarity: having emerged from one fig, they are able to enter another fig by its ostiole, constricted though it may be.[14] In the course of evolution, the agaonids have therefore survived, if I can use the term, the folding over of the inflorescence on itself. The cycle plays out as follows: wasps, both male and female, are born in a fig, and the larvae parasitize flowers. After fertilization, males die, whereas females manage to escape the natal fig (in at least some species, the males, before dying, make a hole for the females in the fig wall). The females, responding to attractive volatile substances, penetrate another fig and lay their eggs in some of the flowers; the parasite presence induces a gall (proliferation of plant tissue) by a process of extended phenotype. As we see, things have barely changed since the ancestors of the figs in that the agaonids have evolved to accommodate the fact that the floral receptacle has folded in upon itself.

What remains to be discussed is the claim that this association is mutualistic. It rests, of course, on the pollination. The female agaonid assures it, carrying the pollen when she leaves the natal fig to lay her eggs in another fig. In certain species, the females, before leaving the natal fig, rub themselves against the male flowers that cover the roof of the fig chamber and therefore become covered with pollen. In other species, a female actively gathers pollen and deposits it in special receptacles on her body to transport it. Having reached the fig in which she will lay her eggs, a female, after laying her eggs in some of the female flowers as I have described, conscientiously spreads the pollen on other flowers; these latter, now fertilized, produce seeds.

The agaonid–fig mutualism is exemplary and gives pause for reflection for at least three reasons.[15] (1) The agaonids cannot reproduce without the figs, while the figs cannot reproduce without help from the agaonids. In particular, the closure of the floral receptacle almost completely prevents wind pollination, and the plant requires an animal able to penetrate the fig. Thus this is a case of obligatory mutualism and is even, I might say, sealed by evolution. (2) The exchange of resources is very clear: the wasp is definitely a parasite of the fig, but the fig, in turn, uses the wasp as a vehicle to spread its genes. (3) The pollination process is particularly economical; in fact, given the specificity of the association, the agaonid that carries the pollen grains is surely going to leave them in the right place, that is, a fig of the same species as the natal fig. Pollen loss will be restricted to whatever risk the insect runs of being rerouted into the mouth of some insectivore. Chemical signals are important in the specificity of these associations, especially as concerns how figs attract wasps. As a rule, pollinators of a given fig species are stimulated by floral scents of this species. Floral blends of different trees always share some compounds, but ratios of these compounds differ among blends (Grison-Pigé, Bessière, and Hossaert-McKey 2002; Grison-Pigé et al. 2002).

I should add that figs, as is true for many fruits, are a gold mine for all sorts of profiteers. Although figs have chemical defenses, they harbor many insect species (wasps, weevils, bugs, butterflies, fruit flies). Many birds and mammals eat ripe figs. Many fig-dwelling insects are predators. Others are parasites, and oddly, many of these are very closely related to agaonids (Rasplus et al. 1998). Thus agaonids that are mutualists (pollinators) and near relatives that are parasites (not pollinators) live side by side. Most parasites lay their eggs in female flowers from the outside, by piercing the wall of the fig with their ovipositors, and galls form there, just as with pollinating agaonids. Some of the parasites, even more economical in their use of energy, are content to lay their eggs in galls previously formed by the embryonic pollinating agaonids. Externally ovipositing parasites cannot play a role in pollination because they have no access to male flowers. However, there are also internally ovipositing parasites (not agaonids) whose females enter the figs and lay eggs in the flowers, just as the "true" pollinator does. This behavior allows young emerging parasitic wasps to carry some pollen from their natal fig to a receptive fig. To an extent, these parasitic wasps help the agaonid that is the "true" mutualist of the fig tree. They provide an interesting clue about the processes by which parasites may shift to become mutualists (Jousselin, Rasplus, and Kjellberg 2001).

Cleaner Fish

Recent research on the behavior of cleaner fish and their clients has elucidated all the subtleties (and fragilities) of this type of association. Cleaner fish include several dozen species from temperate or tropical seas that almost all belong to the family Labridae. Clients of cleaner fish, by contrast, belong to very diverse families. Cleaners and their clients interact at cleaning stations. How do these function?

A cleaning station corresponds to the territory of an individual cleaner. This cleaner rarely leaves the station in which it awaits client fish. In the exceptional circumstance that the cleaner does leave the station, it is to solicit clients, which are not otherwise numerous enough for its taste. If a client fish enters a cleaning station, it generally adopts a particular posture that is primarily characterized by immobility and by what is known as the "pause," to which is added an unusual position, either "head up" or "head down." As soon as the client has thus "declared" its intentions, the labrid cleaner begins to work, swallowing ectoparasites present here and there on the client's skin. The cleaner acts quickly: the labrid swallows twenty times per minute, and most often the inspection and cleaning last several seconds to three minutes.[16] At the end of this time, the client assumes its normal posture and activity, then leaves the station.

The tropical labrid *Labroides dimidiatus* is the best-known cleaner. In the Mediterranean Sea, Céline Arnal and Serge Morand (2001b) have studied another species, *Symphodus melanocercus*. By examining its stomach contents, they have shown that this cleaner is a true specialist in the sense that it rarely eats anything other than what it gleans from the skin of its clients. For example, very rarely did they find benthic animals, that is, individuals picked from the sea floor. By comparing the fidelity of different species of clients to the cleaning stations, they also showed that the benefit drawn by the clients from these visits is truly that they are rid of their ectoparasites, because the visit frequency is positively correlated with degree of parasitism. The parasites are isopods of the family Gnathidae and copepods of the family Caligidae, which are the most frequently eaten.

Earlier I said that this association presents many subtleties, even fragilities. There are several reasons for this qualification. The first is that cleaner fish eat not only ectoparasites but also the mucus of the skin and even some scales. We do not yet know if this consumption of mucus is costly for the client fish, a sort of salary they pay to their cleaners in addition to the parasites they give them. Arnal and Morand (2001a) observe two surprising facts: first, it seems as if, in the previously mentioned cleaner fish *S. melanocercus*, the females consume more ectoparasites than the males, which are drawn more to mucus. Arnal and Morand gave their article a highly metaphoric subtitle: "Females Are More Honest Than Males"! Second, all cleaning visits do not develop in the most perfect harmony. For example, often a client will abandon the pause stage and leave the station before the cleaning is complete. Perhaps this departure results from several maladroit bites by the cleaner, but other explanations are possible.

Other interesting observations come from tropical regions. For instance, when a very large client presents itself, the cleaner *Labroides dimidiatus*, mentioned earlier, does not hesitate to swim right into its mouth and devour parasites in the buccal cavity or branchial cavity. The client, motionless, allows this activity and is careful not to swallow the cleaner. The client can even be a shark several meters long, which remains still, opens its mouth, and lets the cleaner explore its buccal cavity completely calmly (C. Arnal, pers. comm.).

Another surprising fact is that several parasite species seem to have found, by natural selection, a way to parry the destruction wrought by cleaners. They have simply become transparent, or even decorated with spots that allow them to mimic fish skin. These parasites belong to the monogeneans, a group of trematodes studied by Ian Whittington (1996) on the Great Barrier Reef of Australia.

Finally, and this is not surprising in the world of biological evolution, there are cheats—that is, fish with the size, colors, and swimming behavior of true cleaner fish but that are in fact aggressors with sharp teeth who profit from the impunity allowed by this resemblance to approach a client fish nonchalantly and brutally rip off a piece of flesh. Obviously, the cheats cannot be as numerous as the cleaners, or the pause behavior, rendering the clients vulnerable, would be lost to natural selection and replaced by flight behavior, its diametric opposite.

Collaborations between birds known as oxpeckers and the large mammals of the savanna are comparable to those of the cleaner fish and their clients (Mooring and Mundy 1996). Oxpeckers are remarkably effective: the red-billed oxpecker, *Buphagus erythrorhynchus*, can reduce the number of ticks on animals such as the buffalo or rhinoceros by 99 percent. To the list of this type of collaboration should be added the case of the blind snake *Leptophlops dulcis*, which inhabits nests of the nocturnal screech owl *Otus asio* and eats all the insects that appear, thereby greatly reducing the number of ectoparasites that attack the nestlings (Gehlbach and Baldridge 1987). It is the parents themselves who carry the snakes to the nest!

Ménages à Trois

Although we often present associations in the living world as pairwise groupings of species, many obligatory associations entail more than two partners. I will give three examples in which the sheer picturesque quality vies with the instructiveness.

First, the angiosperms (flowering plants) parasitic on other flowering plants are numerous and well known. Some of them attach to the roots of plants; others (like mistletoe) bury their haustoria (attachment organs) in tree branches. It is generally less recognized that there is also a unique case of a gymnosperm (the group of plants including pines and cypresses) that parasitizes another gymnosperm. The parasite, *Parasitaxus ustus,* attains a height of 1–3 meters. The host is *Falcatifolium taxoides,* which can be 15 meters tall. Both are conifers of New Caledonia and belong to the same family, Podocarpaceae.

I include this parasitic association as a mutualism because a mutualist aids the parasite, a bit like the way viruses assist a parasitoid in one of the previous examples. This unique association was described in the 1990s (Cherrier et al. 1992; Woltz et al. 1994), and its parasitic nature is established by several facts. First, it was confirmed that the two gymnosperms belong to two different species by analysis of their chromosomes and certain DNA sequences, and it has even been possible to clarify the phylogenetic relationship of the two species. Also, it was noted that *Parasitaxus* has leaves reduced to scales, has very little chlorophyll, and is never found alone. Finally, it has been shown that *Parasitaxus* attaches itself to the roots or base of the trunk of the host tree by haustorium-like woody root tissue that passes through the bark and spreads in the host cambium layer (the site of plant tissue proliferation).

The most interesting aspect of this association, however, is the presence, in both conifers, of a fungus in the genus *Monochaetia.* The fungus forms mycorrhizae in the roots of the host tree and invades the cells and tissues of both *Parasitaxus* and *Falcatifolium,* without causing visible pathogenic effects in either of them. It appears that the parasitic association between the two podocarps can be sustained only in the presence of the fungus, which carries certain metabolites between the host and the parasite.[17] The fungus is especially indispensable for the germination of *Parasitaxus* seeds, just as the presence of certain fungi is needed for some orchid seeds to germinate.

If these interpretations of the *Parasitaxus–Falcatifolium* association are correct, it is evident that the relationships among the two gymnosperm species and the fungus are complex. The fungus (*a*) confers an advantage on the parasite by facilitating its nutrition and reproduction; (*b*) secures an advantage for the host tree by forming mycorrhizae amid its roots; and (*c*) confers a disadvantage on the host tree by allowing the parasite to establish.

Aside from its originality, this association illustrates the complexity of some associations in which it is not easy to distinguish the costs and benefits for each partner. Here the fungus seems to be a "friend" (a mutualist) of *Falcatifolium* because it helps it to garner nutrients, but a questionable friend because it also allows parasitism by *Parasitaxus*. Perhaps the parasite is the only partner in this triple association to reap only advantages.

A second example is that of carnivorous plants, which are well known. Nowadays, for example, florists try to sell their customers *Drosera* (sundew species), on whose hairy leaves we can see, trapped by viscous droplets, tiny insects that are soon to be digested. In fact, the plant has proteolytic enzymes that attack insect tissues. The products of this digestion are then absorbed by the plant, which benefits from the supplementary organic matter, the rest of which is furnished by its photosynthesis.[18]

An endemic South African plant, *Roridula gorgonias,* a small shrub reaching 2 meters in height, functions differently. Leaves of *R. gorgonias* have many viscous hairs, quite comparable to those of *Drosera*. However, there is an important difference: its viscous secretions have no proteolytic enzyme (Ellis and Midgley 1996). Many insects end up stuck on the leaves of this plant, but what good does it do the plant to capture these insects if it has no means of digesting them? The explanation lies elsewhere. Leaves of *R. gorgonias* harbor many carnivorous bugs of the species *Pameridea roridulae,* found nowhere else. We can also imagine that the mortal traps set for other insects by the leaves are a godsend for the bug. This bug systematically steals prey from *Roridula* and devours them several minutes after their capture. To this point, the bug appears to be a thief who lives at the expense of captures made by the plant on which it resides.

Ellis and Midgley (1996), however, conducted a novel experiment. They "fed" the plant–bug complex with fruit flies they had reared on a culture medium containing labeled nitrogen (^{15}N). Subsequent analysis of the plant showed it had absorbed the labeled nitrogen, even though there was no obvious way it could have digested the fruit fly tissues because it lacked the necessary enzymes. The solution to this enigma is found in a liquid secretion the bugs produce after eating the trapped insects, a secretion the plant absorbs.

The sequence of events is therefore as follows: trapping of insects by the plant; consumption of insects by the mutualistic carnivorous bug; exudation of the liquid secretion by the bug; and absorption of the secretion by the plant.

R. gorgonias is thus a carnivorous plant, but its carnivory is through the intermediary of an auxiliary, a carnivorous insect. Of course, one immediately asks, How is it that the bug is not itself trapped by the viscous plant? This question betrays a lack of confidence in natural selection, which allows the insect to move on the leaves without incurring any risk of getting stuck.

A third example is one that everyone has heard about—the ants and termites that "farm" fungi by constructing gardens in which the fungi grow. These social insects provision the garden with leaves carefully gathered and chewed. For species of *Atta,* South American fungus-gardening ants, this mutualism is complicated by the involvement of a third partner. I illustrate this example from the description by Aron and Passera (2000).

Atta maintain gardens of a basidiomycete fungus, and it is this fungus, because it contains appropriate enzymes, that digests the cellulose of the leaves the ants bring it. Then the ants eat part of the fungus. However, weeds and undesirable fungi can quickly invade such a fungus garden. The ants themselves mount the assault on the weeds by secreting herbicidal substances. As for the attack on undesirable fungi, this is taken on by a third partner, a microorganism of the genus *Streptomyces* (whose antibiotic properties are well known from their use in medicine). These filamentous bacteria are permanently attached to the ant integument. The antibiotics they secrete are especially noteworthy because they destroy a parasitic, inedible fungus that spontaneously grows in these gardens. When the queen ants disperse during their nuptial flights to found new colonies, they carefully carry stocks of these two mutualist partners.

Most remarkable is that the partners of this complex association appear, from analysis of their DNA, to have coevolved for tens of millions of years. It was first demonstrated that the ants and the fungi they culture have coevolved for at least 50 million years, each speciation in one taxon coinciding with a speciation in the other (Curie 2001). More recently, Pennisi (2003) showed that a pathogenic and inedible fungus has probably been integrated into this system from the beginning of this association and has coevolved along with the other two partners.

The Dialogue among Mutualists

There is a group of mutualistic associations of particular ecological importance. These are associations between certain plant species and primitive organisms able to fix atmospheric nitrogen and transform it into ammonia that the plants ultimately transform into nitrates.[19] In fact, no plant (and of course no animal) is able to use nitrogen directly from the atmosphere (of which it comprises nearly 80 percent), although nitrogen is an indispensable constituent of proteins.

For example, the genus *Azolla* contains six species of small aquatic ferns found on all the continents. In small cavities in their leaves, they carry cyanobacteria (sometimes called blue-green algae) related to the genus *Nostoc* and known as *Anabaena azollae,* which fix nitrogen and transfer it to the fern, which thus meets its requirement for this element. *Azolla* is used in many rice paddies in China and Vietnam as "green fertilizer" (after *Azolla* is buried, the assimilable nitrogen is incorporated in the soil). Rice proteins therefore contain atmospheric nitrogen captured by *Anabaena,* transferred to *Azolla,* incorporated in the soil, and finally assimilated by rice.

Casuarina (Australian "pines") are large trees with leaves resembling pine needles, and they have in their roots (as do several other woody plants) nodes containing filamentous bacteria in the genus *Frankia* that can fix nitrogen just as the cyanobacteria in the genus *Azolla* can.

One of the most widespread of all these associations is the one that involves many plants in the legume family (Fabaceae), distributed over the entire planet. These legumes draw nitrogen from their association with soil bacteria that are usually lumped together under the generic name *Rhizobium.* (Different bacterial groups are designated by the global term *Rhizobium* or *rhizobia.* In fact, they are all nitrogen-fixing bacteria only because they share similar plasmids, which bear the genes involved in nitrogen fixation.) Peas, green beans, fava beans, soy, peanuts, but also forage plants such as alfalfa—none of these legumes grow normally unless, during their development, their roots encounter members of *Rhizobium* and associate with them.

We can thus guess that the first reason to be interested in this sort of mutualistic association is the role it plays in feeding humans and livestock. But a second reason is that this is among the associations on which the most research has been conducted, and especially for which the most precise data have been gathered at the level of genes and molecular signals. It is important to recall that, in many parasitic and mutualistic systems I have used as examples, we have seen the results of interactions between genomes, but we rarely understand the details of the mechanisms because we lack adequate experimental data.

For associations between *Rhizobium* and the legumes, however, knowledge is sufficiently advanced that we can largely reconstitute the "molecular dialogue" between two associated species.[20] First, we must recognize that the reduction of nitrogen (N_2) to ammonia (NH_3) is a complex process demanding much energy. An ensemble of enzymes, coded for by about twenty genes (called *nif* genes), allow the nitrogen-fixing bacteria to undertake this reaction. As I noted earlier, no "higher" species contains these genes.

It has been known for more than a century that, when the root of a growing legume encounters *Rhizobium*, the *Rhizobium* are quickly incorporated in an excrescence of the root, called a nodule, which resembles in many ways the galls I have previously discussed. It is in these nodules that nitrogen fixation occurs, and it is here also that a dialogue is established between the partners, in the form of a series of recently elucidated molecular signals: (1) The plant roots exude the first signal, composed of molecules called flavonoids (note that these plant exudates are one of the main sources of soil bacterial nutrition in general). (2) The flavonoids induce the expression of a category of *Rhizobium* genes called *nodD*, which leads to the synthesis of specific proteins. (3) The proteins entrain the activation of a second category of genes, called *nod* structural genes, which leads to the synthesis of other proteins (enzymes) called Nod factors. (Nod factors are not the same in all *Rhizobium* species, which implies a specificity in the plant–*Rhizobium* association;[21] the degree of specificity varies among different environments, along with the diversity of available plants.) (4) The Nod factors leave the bacteria and are recognized by the plasma membrane of the absorbent plant roots. (5) Even in very weak concentrations, the Nod factors then induce a series of changes in the host root (this is called the transduction of the Nod signal). In particular they activate genes in the plant genome, called *ENOD* genes, that induce plant cells to proliferate—this is a new and very beautiful example of the concept of an extended phenotype. The formation of a nodule can be triggered by a single bacterial cell of *Rhizobium*, but these multiply and differentiate quickly to give the nodule interior a bacterial population specialized in fixing nitrogen. (6) Once the *Rhizobium* have multiplied and differentiated in the nodule, nitrogen fixation and the transfer of ammonia to the plant begin.

Much controversy has surrounded the advantages drawn by each partner in this association. For the plant, the advantage is evident: it benefits from the provision of fixed nitrogen. But what is the advantage for the bacteria?

It is important to know that in fact, many bacterial cells of *Rhizobium* live and multiply in the soil without being associated with nodules; they survive, as I mentioned earlier, on root exudates of all sorts (*Rhizobium* are therefore facultative rather than obligatory mutualists). Moreover, it is not certain that the *Rhizobium* that have differentiated in the nodules can reenter the rhizosphere.[22] If this reentry does not occur, then to be closed up in the nodules is the equivalent of suicide for the bacteria (if the bacteria in the nodules cannot transmit their genes, it makes no sense to invoke the usual advantage that people see, that the plant furnishes shelter plus food).

An interesting solution to this apparent paradox was proposed by Olivieri and Frank (1994); they had recourse to kin selection, which we have already encountered in chapter 3 with respect to the trematode *Dicrocoelium*. The idea is that nodule formation allows the host plant to develop and thus to increase its root mass. This increase, in turn, enhances the flow of root exudates, which provide better nutrition for the *Rhizobium* that remain in the soil. If this hypothesis is correct, then to enter a nodule, even to die there, still confers an advantage on the *Rhizobium* population as a whole. This form of altruism can be selected for only if the beneficiaries are, on average, closely related to the individuals that sacrifice themselves. However, whereas Olivieri and Frank think this situation might occur in a local *Rhizobium* population, not all authors agree. Denison (2000, 573), for instance, argues that parasitic mutants should appear in the *Rhizobium* population and be favored by natural selection: "The prevalence of multiple rhizobial strains per plant should favor the evolution of parasitism." Such evolution could be countered by post-infection legume sanctions directed against "cheating" *Rhizobium*. A fascinating series of experiments lends credence to this hypothesis: by artificially preventing normally cooperating *Rhizobium* from actually cooperating (which makes them analogous to cheaters), Kiers et al. (2003) showed that the success of the *Rhizobium* population decreased significantly, which suggests that the plant has some mechanism for punishing cheaters.

In this controversy, it is worth noting that some of the *Rhizobium* seen as mutualists are closely related to other bacteria, in the genus *Agrobacterium*, that are, by contrast, formidable plant parasites.

Agrobacterium tumefasciens induces tumors called crown galls in various plants. Plant cells excised from the gall continue to multiply in culture, and, grafted to a healthy plant, they induce tumor formation. This observation is surprising because the bacterium itself is absent from the excised cells from the gall. The explanation lies in the transfer of a short DNA sequence from the bacterium (a "plasmid" of approximately 200,000 base pairs) to the genome of the plant cell. It is not the bacterium itself that is transferred to the plant cell, only the plasmid. This plasmid inserts itself into the host DNA just as a dancer entering a farandole (I have already used this metaphor at the beginning of this book), and it codes for particular proteins. These, fabricated from materials present in the host cell, are exuded and serve to nourish the *Agrobacterium* living outside the plant. We see that *Agrobacterium* can be considered an ectoparasite that has succeeded, during the course of evolution, in transferring a part of its genetic information into the host genome, to its own advantage. Its interaction with *Rhizobium* constitutes a topic of reflection about the relationship between parasitism and mutualism.

Reciprocal Pressures in Mutualistic Associations

As I briefly argued in the introduction to this book, mutualism differs from parasitism in only one way: instead of one partner in the association exploiting the other without reciprocity, in mutualism each partner exploits the other. Of course, this difference in what we might call the social relationship between the partners has evolutionary consequences. In parasitism, it is in the interest of the victim (host) to rid itself of the aggressor. In mutualism, because each partner exploits the other, there is no benefit to be gleaned for either partner in ridding itself of the other. Must we thus see in mutualism, as many authors have, a peaceful and highly "moral" collaboration that might serve as a model for human society? Not at all. Each partner in a mutualistic association is just as egoistic as those in a parasitic association and has no other goal than to transmit its genes (see the discussion of the controversy about harmony in nature that ends this chapter).

As an example, I will describe a mutualistic association that differs slightly from all the previous ones because it does not imply a truly persistent contact between the partners (it is therefore not a durable interaction in the strict sense of the term). It has, by contrast, the advantage of demonstrating, through an examination of the evolution of a single organ in each partner, how natural selection operates in a mutualism. This is the classic example of some Madagascan orchids and their moth pollinators.

I begin by considering the situation for the orchids. Pollen of many orchid species is gathered in pollinia, which resemble little clubs that secrete a sticky liquid from their bases. Most of the flowers also have nectaries, at the bottom of which is a sugary liquid. A moth attracted by the odor of the flower inserts its proboscis into the nectary. To reach the nectar, it rubs against the bases of the pollinia, which stick to its head. Thanks to their successive nectar meals in different flowers, the moths transport the pollinia and assure cross-fertilization, indispensable for orchid reproduction. However, for the pollinia to stick to the moth, the moth must move its head sufficiently far and with sufficient force into the corolla. If the moth does not enter completely into the flower, it takes the nectar but leaves without the pollinia. Therefore, the only flowers to exchange pollen with other flowers are those whose nectaries are located at the end of a long, tubular "spur," forcing the insect to go to some length to reach the nectar. If there is genetic variation among individual plants in length of nectary-bearing spurs, selection favors the trait of "long spur."

Now I examine this case from the standpoint of the moth. If a moth with a short proboscis tries to drink nectar from a flower with a long spur, it will be unable to reach the precious liquid. This can happen to this poor individual with many flowers. Fatigued and malnourished, this moth will have diminished or no reproductive success. Moths with long proboscises are thus favored, and, so long as there is genetic variation for proboscis length, genes coding for the trait "long proboscis" will increase in frequency in the moth population.

Finally, I examine things from the perspective of the association. If the trait "long spur" has a selective advantage in the orchid population and the trait "long proboscis" has a selective advantage in the moth population, the resulting process leads to a continuous lengthening of both spurs and proboscises. The reciprocal pressures (I will discuss coevolution in chap. 6) continue in a remarkable manner, "creating" orchids with outlandishly long spurs and moths with correspondingly long proboscises. To cite just one example, nectaries of the orchid *Angraecum sesquipedale* are on spurs from 28 to 32 centimeters long, and the moth *Xanthopan morgani* that pollinates it has a proboscis of more than 25 centimeters! Remarkably, Darwin knew of the orchid and predicted that there must exist on Madagascar a moth with a proboscis as long as its spurs. In the 1920s, the moth was discovered and Darwin was proved right (Wasserthal 1997; Nilsson 1998).

We see that (1) the fact that the association is mutualistic rather than parasitic does not preclude reciprocal selective pressures, and (2) the collaboration between the moth and the orchid is no altruistic process, even if it has the appearance of a perfect entente. Moreover, defenses exist even within a mutualistic system. They are not used to destroy the other partner but to prevent the egoism of the other partner from leading the entire mutualism to harmful excess. For example, if an insect harbors bacteria that provide it with nutrients, it is still necessary that these bacteria, while multiplying, do not invade the body of their host. Therefore selection arises in the insect genome for mechanisms that assess and control bacterial multiplication.

To complete the example of orchids and their pollinators, I devote the section that follows to an overview of what might be called the hidden egoism of mutualisms.

Controversy: Harmony in Nature

Snow White and Bambi

Nature presents to humans a picture of harmony, at least in some circumstances. Forest animals are the best example: in cartoons, they always come to the aid of humans maltreated by other humans (the birds that succor Snow White) or even to the aid of each other (Bambi's companions). Even earthly paradise, as described in the Bible, has the predator lying down peacefully with its prey.

Some examples of mutualism that I have cited in this chapter can lead one to believe in the evolution of this sort of harmony. However, one can also defend the diametrically opposite viewpoint, that harmony is nothing but an illusion: everyone knows perfectly well that spiders chase innocent insects in the understory and on the ground, that snakes smother little rodents, and that foxes eat birds.

If I add to this unappetizing vision the sorts of things that go on in the world of parasites, then the picture becomes totally bleak. For the biosphere has many more parasitic beings than free-living ones. Wherever we look, infective stages seek out their targets, suckers and hooks only waiting for the chance to latch on, enzymes ready to flow onto the anticipated host.

I begin this controversy over the question of harmony by taking a step back—that is, by beginning with interspecific interactions and moving toward the DNA molecule, bearer of the information that determines everything.

Harmony between Species?

The examples of mutualism that I cite in this chapter should suffice to convince readers that interspecific mutualistic associations play a key role in the lives of the interacting species. Although we can imagine that client fish might be able to do without their cleaner fish and nematodes without their bacteria, we can clearly see that the fig can no longer live without its wasp and vice versa. But what role do such associations play overall at the level of the biotic community?

Leigh and Rowell (1995) believe that mutualism has led to at least three advances: the colonization of certain inhospitable environments, an increase in biodiversity, and the tremendous proliferation of certain groups in the course of evolution.

The conquest of inhospitable environments is illustrated by the mutualism between plants and mycorrhizal fungi and that between corals and dinoflagellates.

Many vascular plants (more than 70 percent) are associated with soil fungi known as mycorrhizae (see Smith and Read 1997). The mycelium of the fungus surrounds the radicels (tiny secondary roots) of the host plant, insinuating itself between the cells, or even forms branched masses within cells. It receives nutrients in the form of carbohydrates from the host plant. In exchange, the fine fungal network "explores" the soil better and over a larger area than do the roots of the plant itself, and the fungus brings the plant supplementary minerals. The majority of plants that are usually mycorrhizal can grow in the absence of mycorrhizae, but such individuals are stunted. The poorer the soils, the more crucial is the advantage conferred by the mycorrhizae. Sometimes, the conquest of a particular environment is possible only to species associated with a suitable fungus. Such is the case, for example, for heathers, which would be unable to grow on acid soil without their mycorrhizae.

As for the corals of tropical seas, they are associated with unicellular photosynthetic organisms, the dinoflagellates, often called zooxanthellae, that are found in their cells. Surprisingly, there can be several species of zooxanthellae in one coral species, as has been demonstrated by DNA analysis. The energy that these endocellular mutualists bring to this association has allowed many coral species to flourish in nutrient-poor waters.

Increase in biodiversity is classically illustrated by tropical forests characterized by many tree species, each of which is usually represented by just a few individuals separated from one another by considerable distances. The agaonid–fig mutualism shows that pollen transport by highly specialized insects provides the best guarantee that pollination will be successful, even if density of fig trees is low. In other words, it is possible for rare species to persist. Because very many tropical forest plant species use animals either as pollinators or seed dispersers, each species can be rare and the number of species is thus high. Of course there is a limit to rarity; for example, the lifespan of a fig's wasps and the wasps' flying abilities mandate a limit on the distance between figs of any given species (which sometimes can be as much as 10 kilometers).

The tremendous proliferation of some groups in the course of evolution poses a difficult question. What can explain the explosive diversification of flowering plants that began during the Mesozoic? Today there are many more angiosperm than gymnosperm species. Some attempts to explain this disparity refer to differences such as those found in vascularization and growth rates (Bond 1989; Gorelick 2001). But it is also possible that some increases in diversity were made possible by associations, ephemeral or persistent, among various organisms. For plants, it is often suggested that associations of angiosperms with insect pollinators, on the one hand, and with vertebrates dispersing the seeds on the other, gave angiosperms an advantage over gymnosperms. The latter, for instance, with the exception of Cycadales, retain wind pollination. It is worth adding that the evolutionary radiation of angiosperms allowed selection for new mutualisms. An example is defensive mutualisms in which insects defend plants against other, plant-eating insects (phytophages). Discussing preadaptation in chapter 2, I cited ants that inhabit domatia of certain plants and, in exchange, constantly patrol the host plant and protect it against insects that come to devour its leaves.[23]

We should conclude that, at the community level, mutualisms of all sorts have played and continue to play an important role in nature. It is important to realize that this observation in no way contradicts the fact that mutualistic activities mask the egoism of genomes. Moreover, there exist cheats in the majority of associations. There are countless examples, such as the false cleaner fish (which mimics cleaners and profits from its appearance, as I have just described, to get close to clients and bite off pieces of their flesh), the plant that causes itself to be pollinated by insects without offering them a nectar reward (the labellum of certain orchids even mimics a female wasp and thus attracts male wasps), or the ants that expel mutualistic ants from their domatia but do not repel plant-eating insects. A list of cheats would be endless.

Harmony within Species?

Associations between individuals of the same species (conspecifics) are normal. They are more or less loose and persistent, running the gamut from unstructured schools of fish to highly organized insect societies. And of course I must add human society, with its complexity, its endless forms of association, and its innumerable forms of cheating (see, for example, Klein 1989; Bowles and Gintis 2002; Mesoudi, Whiten, and Laland 2004).

Just as a parasite and its host are in competition for the same resource, individuals comprising a population are in conflict for space, food, and reproduction. And just as relations between a parasite and its host can take the form of mutualism if the interest of each partner is thus served, so individuals of the same species can be induced to cooperate.

Associations within species arise from three distinct processes.

The immediate advantage derives from group living without a particular group structure. For example, a gregarious tendency can be selected if the probability of being eaten by a predator is greater for a solitary individual than for a group member. One possible explanation for this advantage could be that an alarm raised by one member benefits all others in the group. In a study of Australian forests, Serge Aron and Luc Passera (2000) report that solitary bird species are confined to the poor environment of high branches while gregarious species can exploit the rich environment of the ground, despite the presence of predators, and this exploitation is a benefit of alarm calls. In other instances, the advantage of group living may be the ability to capture large prey, which could be impossible for a lone predator.

A deferred advantage is represented by what is known as *reciprocal altruism,* which requires the cognitive faculties necessary for individual recognition. Pierre Jaisson (1993) cites as examples the olive baboon and the Azara vampire bat. Among the baboons, a young male has a much greater chance of fertilizing a female if he is helped by another male who stands guard, preventing any competitor from approaching during copulation. Later, this "altruistic" male will solicit assistance from his beneficiary in a similar circumstance. As for the Azara vampires, individuals that are well fed during the night regurgitate blood into the mouths of vampires that had not, during that night, found victims. Of course we are not surprised to find once again the possibility of cheating in reciprocal altruism: the individual that accepts a service but refuses to render it has a selective advantage because this individual receives a benefit without paying a cost (unless natural selection has produced mechanisms to punish cheaters).

Finally, kin selection explains the evolution of the full-fledged ant, bee, and termite societies. More than any other type of persistent association, these insect societies give the impression of true harmony. In fact, this harmony is achieved to the detriment of certain individuals, because it is founded on the existence of "castes," and certain castes have no other way to transmit their genes than to "serve" reproductive kings and queens. An entire branch of biology (sociobiology) is based on the idea that the reproductive success of an individual should be measured not solely by the success of its own genome but also by the success of genes that are identical to this individual's genes and are present in genomes of relatives. In working for the queen, who is their sister, the workers of a beehive are working for the success of their own genes. This type of selection is called kin selection. I discussed it in chapter 3 with respect to the metacercariae of *Dicrocoelium* and in the present chapter with respect to *Rhizobium*.

Whatever the type of mutualism at play in intraspecific associations, the apparent harmony results from convergent interests and is not a form of altruism. The expression *reciprocal altruism* even seems to be ironic because it describes nothing but the results of the pure self-interest of the supposed altruists.

Harmony among Organelles?

The greatest success of cooperation among organisms with initially distinct genomes is that of the eukaryote cell.[24]

Organelles such as mitochondria that inhabit each of our cells have a characteristic that appeared surprising for a long time: they have their own DNA, distinct from that of the cell nucleus, and their own DNA codes for molecules of RNA and proteins.[25] The amount of DNA in a human mitochondrion is small compared with that in the nucleus (slightly more than 16,000 pairs of nucleotides as against 3 billion), but this mitochondrial genetic information is indispensable to cellular metabolism. The chloroplasts of plant cells contain 120,000 to 150,000 nucleotide pairs, depending on the species.

The fact that DNA exists outside the nucleus was long interpreted as the result of migration of that DNA from the nucleus. However, today we know that this migration is not the origin of this DNA. Mitochondria (in animals and plants), chloroplasts (in plants only), and probably also several other cellular organelles are nothing other than bacteria that have become mutualists, perfectly integrated into the eukaryote cell (see Selosse and Loiseau-de Goër 1997; Selosse, Albert, and Godelle 2001). At first this hypothesis seemed bold and rested only on facts such as the general appearance, size, and mode of division of mitochondria and chloroplasts, all of which resemble those of bacteria. The biologist Lynn Margulis (1992) accumulated arguments in favor of an "exogenous" origin of organelles, and comparing the analysis of bacterial DNA with that of the eukaryote cellular organelles provided definitive proof that our cells result, without exception, from a mutualistic association believed to be at least a billion years old.

P. W. Price (1989) thinks that the evolution of eukaryotes, the acquisition of organelles, and the evolution of biotic complexity have depended on the initial invasion of parasitic organisms, an invasion that became mutualistic secondarily. It is, in fact, highly likely that at some moment in evolution, bacteria penetrated living cells and survived there. These bacteria were then transmitted indefinitely, dividing as the host cells divided. This is therefore an ancient association, without which evolution would certainly not have led to the exceptional successes (humans included) that followed.

It is easy to understand why organelles associated with eukaryotic cells do not behave like parasites. They are in a situation in which there are no new hosts to infect because vertical transmission assures that all eukaryotic cells receive organelles.[26]

Is the association between eukaryotic cells and mitochondria perfectly harmonious? Not at all! It is important to know that mitochondrial transmission occurs only or almost exclusively through female gametes,[27] and that this mechanism has two opposing consequences.

The first is that this transmission mode limits the possibilities of conflict between mitochondria themselves at the moment of fertilization. If the ovule and spermatozoid each carried its mitochondrial stock, they could differ genetically and enter into conflict, which could only harm the host cell. The fact that, in general, the zygote receives all its mitochondria from the ovule eliminates the possibility of conflict. This reasoning applies to uniparental transmission in general (see Godelle and Reboud 1995).

The second consequence, by contrast, is that this transmission mode generates a conflict between the mitochondria and the host cells themselves. Because the mitochondria are transmitted only by females, it is in their "interest" to favor production of females. In fact, there exist mitochondrial mutants able to bias sex ratio in favor of females. Hermaphroditic individuals and individuals whose male function is sterilized coexist in natural populations of some plants.[28] The male sterility results from the influence of mitochondria that favor their own transmission. This distortion of the sex ratio to favor females is counterbalanced by the action of nuclear genes that restore a balanced sex ratio.

The intimate association between the nucleus of the eukaryote cell and the mitochondria and chloroplasts, maintained over a very long period, has allowed the bearers of information—that is, the DNA molecules—to be themselves the object of transfers. It has been proven that genes present in the nuclear genome of the eukaryote cell have come from mitochondria and, in plants, from both mitochondria and chloroplasts. It is estimated that the mitochondria have lost 99 percent of their initial bacterial genome, and part of this loss is accounted for by transfer to the nucleus. It is not impossible that other cytoplasmic structures of modern eukaryote cells, such as peroxisomes and centrioles, are also ancient bacteria that have lost all their DNA to migration to the nucleus, thus achieving a complete transfer of information between two species that were initially distinct. It appears that different selective pressures favor, on the one hand, a concentration of the genes in the nucleus and, on the other hand, a reduction in the size of the genomes of bacteria that have become mutualists.

Several biologists, especially the American Carl Woese, believe that gene transfers, analogous in principle to those I have just described, occurred more freely and frequently toward the beginning of the history of life. Woese thinks we should seek not a single cellular root for the tree of life but rather a sort of cellular network whose genomes are enriched by frequent exchange of sequences. Later this network gave rise to the three main branches of the tree of life recognized today: the eubacteria, the Archaea, and the eukaryotes. Some researchers even think there might have existed a fourth branch, now extinct, but which has left a legacy in modern genomes in the form of genes acquired in ancient transfers. Woese (2002) has written that horizontal gene transfer is "the principal driving force in early cellular evolution."

Harmony between Genes?

"It is difficult to admit that a conflict can exist between constituents of the same organism," wrote Gouyon, Henry, and Arnould (1997, 209). However, as these authors quickly recall, this type of conflict certainly exists.

Sexual reproduction is based on meiosis and fertilization. At meiosis, the chromosomes are apportioned by chance between the daughter cells such that the two alleles of the same locus should normally be "honestly" divided among the gametes.[29] If locus X is in a heterozygous state (that is, occupied by two different alleles, a and b), 50 percent of the gametes should carry a and 50 percent should carry b.

If there really exists a realm in which one might believe that it is unnecessary to state that mutualism will reign, it should be among the genes of a single individual. Meiosis should be honest. But sometimes it is not!

Two alleles can enter into conflict in the sense that one of them can induce a distortion of meiosis in its own favor (see the controversy in chap. 2 over strange parasitism). Such egoistic alleles exist and are beginning to be recognized in many species. There are also entire chromosomes that bias meiosis in their favor by inducing the destruction of the homologous chromosome. Of course such an occurrence is catastrophic for the genome as a whole, and antidistortion mechanisms exist. It is nevertheless striking that genomes have to protect themselves against some of their own components. Leigh (1971) uses an elegant metaphor—that chromosomes constitute a "parliament of genes" in which most members can unite for the general good to oppose the cabals of a few troublemakers.

The important conclusion, however, is that troublemakers exist everywhere, and even parts of living organisms act as troublemakers when it is in their own interest.

Alice and the Red Queen

Coevolution

What Is Coevolution?

Because biological evolution is generated by a permanent confrontation between genetic variation and selective pressures, every living species can constitute a source of selective pressures for other species. For example, by running after gazelles in order to eat them, cheetahs exert selective pressure on gazelle populations, because, in a statistical sense, they capture the individuals least able to escape. This selection favors gazelles that run fastest, that most quickly detect the presence of danger, and that make good decisions to evade their pursuers.

In fact, this is a two-way and not a one-way process. For even if we do not always think of it at first, the gazelles also exert selective pressure on cheetahs. If cheetahs have genetic diversity for traits such as ability to detect gazelle herds or speed of pursuing them, individuals that are most adept in these matters will be the best fed. Now, those that eat best will also be those leaving the most descendants (the males because they will copulate with the best females, the females because they will best defend their cubs), so genes for good detection or great running speed will increase in frequency each generation in the cheetah population.[1]

What I have just described is the phenomenon of coevolution. Coevolution is the process by which two adversaries never cease acquiring new adaptations so that neither "gets ahead" of the

The change of the living world after the origin of life on earth is designated by the simple word *evolution*.

◊

ERNST MAYR (1999)

other. There is a concatenation of reciprocal selective pressures (that is, of cheetahs on gazelles and gazelles on cheetahs). I have already described the example of the orchids and their moth pollinators (chap. 5).

The example of cheetahs and gazelles, however, is a bit too much of a caricature for the simple reason that cheetahs are not the gazelles' only enemy and gazelles are not the cheetahs' only food. There are certainly reciprocal selective pressures between gazelles and cheetahs, but they are diluted in a complex mix of conflicting selective pressures exerted on one another by the many species inhabiting a savanna ecosystem. In order for the concept of coevolution to keep its original meaning, the term *coevolution* (sensu stricto) must be reserved for confrontations in which the number of species generating reciprocal selective pressures is either (ideally) reduced to just two or at least limited to several well-identified adversaries, and we should speak of *diffuse coevolution* when the number of interacting species is greater.

With this definition, persistent interactions are the prime domain of coevolution because of the typically narrow specificity of parasitism. Although the great majority of hosts harbor more than one parasite species, many parasite species, on the contrary, have only one host species or a very small number of them. Parasitic birds are probably one of the best examples to illustrate the principle of coevolution.

The Life of Cuckoos

In fourth century BC, Aristotle discussed the biology of the common cuckoo, today known as *Cuculus canorus,* more or less in these terms: "It lays its eggs in the nests of small birds and does not rear its young; when these are born, they throw the other chicks present in the nest over the edge." And at the end of the eighteenth century, the English biologist Gilbert White called such behavior a "monstrous outrage perpetrated on maternal affection."

The biology not only of the common cuckoo but also of other parasitic birds has been well studied because it raises questions that fascinate biologists, especially evolutionists.

I should clarify at the outset that there are two sorts of parasitism among birds. One is occasional or facultative, the other obligatory. Occasional parasitism is especially seen in birds that live in dense colonies, such as the starling (*Sturnus vulgaris*), but also the barn swallow (*Hirundo rustica*), and even the common moorhen (*Gallinula chloropus*). As noted by N. B. Davies (1988), many of these birds "play at cuckoo." This is a matter of intraspecific parasitism: simply stated, a female takes advantage of the absence of a neighboring couple from their nest to lay an egg in it.

Of course one might object that this behavior is only a "mistake." However, the fact that, in some species, it is an egg already partly incubated that the parents carry from their nest to the nest of a neighbor proves that, at least in those instances, it is an authentic parasitic act. Some of these birds (starlings, for example) have even been seen to evict an egg from the host nest before depositing their own, an act that can only improve the average quality of care each of their own young receives from the involuntarily adoptive parents.

Defense mechanisms against intraspecific parasitism appear to have been selected in these species. One of them, called *parasitism insurance* (Rothstein 1990), consists in laying fewer eggs than what would seem to be optimal (that is, the female lays fewer eggs than the number of nestlings she can feed). In this way, if extra eggs are laid in the nest by a strange female, the original clutch is not at risk of malnourishment. In intraspecific parasitism, the recognition of "foreign" eggs is a priori difficult because the eggs of the exploiter and the exploited birds are very similar. Nevertheless, Lyon (2003) has shown that recognizing one's own eggs even among others of the same species is possible for certain birds such as the American coot, *Fulica americana*. Lyon suggests that the female of this species is able to count the eggs in its clutch and therefore to detect fraud.

In all cases mentioned so far, parasitism is facultative, and the tricked parents nevertheless rear at least some of their own young. To return to N. B. Davies's metaphor, the interlopers are not really "professional parasites."

Such is not the case for other parasitic birds, which are, on the contrary, formidable professionals. Obligatory parasitism is seen in about 1 percent of all bird species, belonging to several families (Winfree 1999). The common cuckoo, distributed in Europe and Asia, is the classic example. Its biology can be summarized as follows.[2]

Cuckoos do not build nests. Upon returning from their winter migration to the heart of Africa (French cuckoos go to the region between Mozambique and Lake Tanganyika), the females, after having been fertilized (often by several males), lay their eggs in the nests of various bird species. The female cuckoo capitalizes on the temporary absence of a female passerine that has begun laying her eggs to land on her nest, swallow one of the eggs present, and replace it with one of her own. Often, this whole maneuver takes no more than 20 seconds. The female cuckoo then victimizes another nest, and another, and so forth, so that the eggs of a single female (up to 20) are dispersed among nests of several adoptive parents. From then on, neither the male nor the female cuckoo lavishes any care whatsoever on their offspring. Unless this fraud is detected by the adoptive parents, they incubate the cuckoo egg along with their own eggs. The cuckoo egg nearly always hatches first (about 48 hours before the eggs of the host bird). As Aristotle observed, the young cuckoo, even though it barely weighs 3 grams and is still blind, takes this treachery so far as to toss the host eggs out of the nest. It achieves this operation thanks to two remarkable adaptations. On the one hand, the young cuckoo is able to "grasp" the host eggs betweens its still embryonic wings; on the other hand, it has a small, rounded depression between its two shoulders where it places the egg. From then on, it is only a matter of carrying it to the edge of the nest. In the exceptional case that host eggs have hatched first, it is the chicks themselves that the young cuckoo throws out.

At that point, the adoptive parents have no one in the nest but the young cuckoo, and they feed it royally. This behavior is all the more remarkable because the young cuckoo does not resemble their own nestlings at all. After two or three weeks, the parasitized passerines end up feeding chicks much heavier than they are; a three-week-old cuckoo weighs close to 100 grams, whereas most adult passerines parasitized by cuckoos weigh from 10 grams (for example, the wren *Troglodytes troglodytes*) to 30 grams (for example, the great reed warbler *Acrocephalus arundinaceus*). The young cuckoo grows quickly to adulthood without ever having known its biological parents.

Coevolution in the Relationship between the Cuckoo and Its Host

For any host bird (I will use a reed warbler as an example), the cost of being parasitized by the common cuckoo is enormous, because the reproductive success of the pair falls to zero for the breeding period under consideration. This drop constitutes a substantial selection pressure, because, within the warbler population, if some individuals have genes that enable them to thwart the cuckoos, these individuals will be selected for and their genes will spread. As it happens, such genes do exist.

If we artificially manipulate a clutch of host eggs by replacing one egg with a plaster egg, we notice that, upon returning to the nest, the proprietor usually perceives the attempted deceit if the plaster model mimics the egg only superficially. The bird is then able, in many instances, to rid the nest of the model by tossing it over the edge. Only if the model resembles the egg closely (size, spots, reflectance) is the host tricked.

If we now consider the situation in nature, in which warblers and cuckoos confront each other generation after generation, we hypothesize that, if certain individuals in the warbler population are better able than others to discriminate cuckoo eggs from their own eggs, natural selection will favor the genes of these discriminating birds. This is an example of stage 1 of coevolution (fig. 6.1). The behavior of eliminating the foreign egg, called rejection behavior, then constitutes a selection pressure, but this time a pressure on the cuckoo population. We readily guess that this pressure has led to the evolution in cuckoos of individuals able to lay eggs more closely resembling those of their host species. This is stage 2 of coevolution.

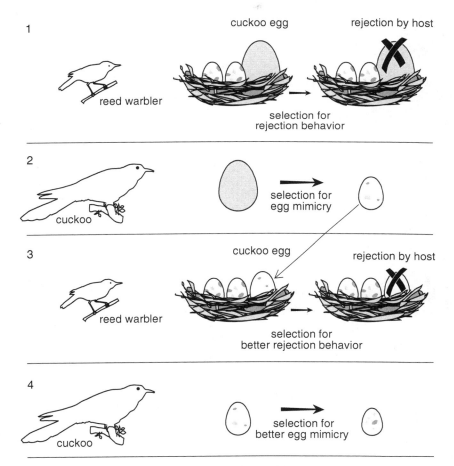

Fig. 6.1. Diagram of an arms race in the cuckoo–host bird system. Natural selection retains the cuckoo eggs that are increasingly similar to those of the host species; it also retains, in the host, the increasingly effective behaviors of rejecting foreign eggs. (Note that the selection process occurs over the course of many generations and not just in the four shortened stages shown.)

To this increased mimicking ability of the cuckoos, natural selection in the host species should respond by producing even stronger powers of discrimination, leading to rejection of the foreign egg even if the latter shares many traits with host eggs. This is stage 3 of coevolution. We can easily guess the next stage: still more accurate mimicry by cuckoo eggs of host eggs, obtained by natural selection among the various eggs laid by the cuckoo population. This is stage 4 of coevolution.

The following stages are only an indefinite reprise of the preceding ones. Among cuckoos, there is selection for eggs ever more similar morphologically to those of the host. Among hosts, there is selection for ever-greater powers to discriminate foreign objects in the nest. Of course this selective process occurs only if there is continued renewal of genetic diversity in populations of both adversaries, generated by mutations or by recombination of genes allowed by sexuality.

There are two strong arguments for believing that the coevolutionary scheme I have just sketched corresponds to reality, at least in outline. The first is that the cuckoo egg is remarkably small given the size of the adult cuckoo, and it truly resembles the eggs of the parasitized birds in many traits. In some cases, it is not easy even for a skilled ornithologist to distinguish cuckoo eggs from those of its host. The second is that in several fortuitous cases, there is evidence over a period of several decades of selection for rejection behavior in bird species newly confronting certain introduced parasitic bird species (Robert and Sorci 1999).

Coevolution between cuckoos and their hosts constitutes an exemplary case of an evolutionary "arms race," a metaphor that refers to two nations in conflict, each of which is forced to acquire better armaments each time its adversary acquires new weapons (see the definition in chap. 1). When we examine details, however, things are a bit more complicated than I have described them.

The most important additional fact is that common cuckoos are not specific to a given host species. For example, in France, they currently parasitize ten passerine species (to the great reed warbler must be added, for example, pipits, wagtails, robins, redstarts, and dunnocks). Now, each of these species lays its own distinct kind of egg, and it is not possible for a female cuckoo to "decide," according to the nest in which she will lay her egg, whether to give it a particular color or spotting pattern. In fact, it appears that an individual female cuckoo lays eggs of the same type throughout her life and so can successfully parasitize only one host species. The differences in egg appearance among different cuckoo females are so pronounced that, at one time, experts thought there were as many cryptic cuckoo species (that is, almost indistinguishable morphologically) as there were host species. Genomic analyses prove that this is not the case and that male cuckoos copulate indiscriminately with females specialized for different host species (Marchetti, Nakamura, and Lisle Gibbs 1998). The genetic mixing arising from sexual reproduction thus opposes any specialization in the cuckoo population as a whole for parasitism of one particular host species. All cuckoo females exploiting a particular host species constitute a *gens* (plural, *gentes*).

We would expect traits involved in egg mimicry to be inherited only from female cuckoos, so that the genes coding for these traits would by transmitted only by females. Such inheritance is possible if the genes are located on the sex chromosome determining femaleness or if the genes are part of the mitochondrial DNA.[3] Lindholm (1999, 294) writes that "such sex-linkage would allow the preservation down the matriline of favourable combinations of genes for egg mimicry, regardless of the male genotype." However, neither possibility is likely. It is noteworthy that the cuckoo is probably the only parasite that has succeeded in specializing in a series of different hosts without itself splitting into several species.

One extra difficulty in the coevolution of cuckoos with their hosts is that the hosts have usually not developed defenses against the young cuckoo itself, if only by not feeding this huge nestling that does not resemble their own nestlings at all. Possibly the explanation for this failure lies in the fact that any defense entails a cost. The cuckoo's throat is bright red, and this brilliant mark in the large open beak stimulates feeding behavior by the adoptive parents. Also, the cry of the young cuckoo more or less mimics the squawking of an entire clutch. Now, these are the same stimuli that elicit the normal behavior of the hosts feeding their own young. It may be, therefore, that if the hosts were to stop responding to these stimuli, matters would be worse (in terms of not feeding their own young) than if they continue responding.

In fact, recent research shows that, for at least one cuckoo species, the arms race has gone further than simply detecting a suspicious egg. This species is the superb fairy-wren (*Malurus cyaneus*) of Australia, which is parasitized by Horsfield's bronze cuckoos (*Chrysococcyx basalis*). In this association, as in that of the common European cuckoo, the cuckoo nestling evicts all host eggs or young soon after it hatches and remains alone in the nest. If the egg-laying periods of the host bird and the parasite coincide exactly, the female superb fairy-wren manifests no rejection behavior. The eggs of the cuckoo are very similar to those of its host, but experiments show that the fairy-wrens do not even reject artificial eggs that are very different from their own. Defenses against eggs are almost totally lacking.

Langmore, Hunt, and Kilner (2003), however, show that even though fairy-wrens cannot recognize eggs of Horsfield's bronze cuckoos, they have evolved to discriminate against cuckoo chicks. In slightly more than a third of observed nests, female fairy-wrens abandoned the nest and its cuckoo chick, which quickly died of hunger and was eaten by ants. The female fairy-wren did not even wait for the cuckoo chick to die before constructing a new nest and proceeding to produce a new clutch.

It appears that the biology of superb fairy-wrens favored selection of abandonment of parasitized nests. In fact, this host species can produce up to three clutches per year. Therefore, when a female deserts the nest with the cuckoo chick, she can quickly produce a replacement clutch. The cost of the risk of rejecting its own offspring is thus minimized.

How does the female superb fairy-wren recognize the cuckoo chick? Neither by the fact that it is alone in the nest nor by its shape or color, it seems. Rather, it recognizes the vocalizations of the cuckoo chick. Interestingly, all indications are that Horsfield's bronze cuckoos have been selected to produce vocalizations very similar to those of the host chicks, which constitutes yet a new stage in the arms race!

To give a more complete account of cuckoo biology, I should add that the possible defenses of the host birds are not limited to the rejection of intruder eggs or nestlings described above. For example, some birds nesting near the ground, such as the rufous scrub-robin (*Cercotrichas galactotes*), have learned (by natural selection, that is) to nest as far away as possible from trees (Alvarez 1993). Because cuckoos always perch on a high tree to locate nests of their potential victims and to observe their comings and goings, to nest far from trees is an effective countermeasure.

Conversely, there are passerines that, for reasons not yet clear, have practically no defenses against cuckoos. Such is the European dunnock (*Prunella modularis*), which is termed an "acceptor" bird because it appears incapable of any rejection behavior, even if the cuckoo eggs or other objects placed in its nest in no way resemble its own eggs.

As noted earlier, the common cuckoo is hardly the only parasitic bird species. In Europe, for example, there is another species, the great spotted cuckoo (*Clamator glandarius*), which lays its eggs in the nests of magpies and related species. Its biology differs a bit from that of the common cuckoo, notably in that the young great spotted cuckoo does not eject the host eggs, so the magpies raise their own young and the young cuckoo side by side. In the great spotted cuckoo–magpie system is found what is perhaps the ultimate cuckoo weapon, which has been called "Mafioso behavior." Cuckoos have been observed to punish magpies that reject cuckoo eggs, by returning to their nests and stamping on and destroying their eggs. Thus, the only magpies that transmit their genes are those that do not demonstrate rejection behavior; the details of how this behavior evolved are still unknown.

Among parasitic birds that are unrelated to cuckoos are the famous New World brown-headed cowbirds (*Molothrus ater*). In the United States, these cowbirds form flocks of millions of individuals and constitute a grave threat to several other bird species. The young cowbird does not attack the nestlings of its host, but the nest can come to have more cowbird eggs than host eggs, so that host reproductive success is greatly lowered (and we should not forget that the host clutch has other enemies, especially predators). The considerable impact of cowbirds is proven by many observations. For example, on a California military base, the number of pairs of a small endangered passerine, Bell's vireo (*Vireo bellii*), grew from 20 to more than 250 when cowbirds were destroyed. Entire scientific meetings have been devoted to research on ways to limit cowbird impact. This impact is all the greater because cowbirds are generalists that lay their eggs (not mimetic) in nests of many host species (220 according to Rothstein and Robinson 1994). Human-caused ecosystem fragmentation (from deforestation, road construction, etc.) allows cowbirds to invade areas where they had formerly been absent and to attack new host species in which natural selection has not had time to generate effective defensive behaviors.

Of course the fact that defensive behaviors are sometimes selected for and other times not, are sometimes effective and sometimes inadequate, is not easy to explain. Doubtless it requires consideration of several ecological or evolutionary factors, such as the relative density of parasitic birds or how old the parasitic association is. For example, it has been shown that rejection behavior cannot be selected for if there exists a substantial risk that the host will erroneously eject its own eggs rather than those of the parasite. The age of association also plays a role insofar as there is always a lag between a selective pressure and its evolutionary response. Just as humans initially have no defenses against emerging pathogens, so many generations can pass before a bird species newly parasitized by cuckoos can evolve a rejection behavior. For the dunnock, for example, it has been suggested that its passive acceptance behavior may signify that it was not a host of cuckoos until recently (perhaps around the fourteenth century).

Anticuckoo Defense and Pair Fidelity

Analysis of behavioral relations between the reed warbler *Acrocephalus scirpaceus* and the common cuckoo *Cuculus canorus* led Davies et al. (2003) to astounding conclusions. These authors show that even a glimpse of cuckoos can lead reed warblers to desert a "parasitized" nest but more often merely causes increased surveillance of the nest by the host birds. However, this increased surveillance is carried out entirely by the male. The female, probably because of constraints associated with searching for food, does not modify her surveillance activities.

Now, the changed male behavior presents a danger, which is that the female, who continues periodically to fly far from the nest, can be fertilized by another male. Davies et al. (2003, 288) observe that infidelity is not rare among birds. They note that, to limit the risk of female infidelity, male reed warblers change their surveillance behavior only after the female has laid at least one egg: "males spent most of their time close to their females both during nest building and up to the laying of the first egg" and "followed the females much less once the first egg appeared." The authors show that there is nevertheless a small difficulty in that, because the eggs are fertilized 24 hours before laying, the males seem to relax their attention to the female a bit too early. They suggest that in fact the danger of infidelity is lower at this moment because the females, having laid the first egg, do not as readily seek extra-pair matings.

These results show that the lives of certain birds are indeed difficult because they must choose between the risk of parasitism and the risk of infidelity by their mate. The fact that males cease watching the females closely as soon as egg laying begins also shows that they are defending not their honor but simply the transmission of their genes.

To finish with cuckoos, I will ask whether the young cuckoo receives parasites such as feather lice or feather mites from its adoptive parents. For the common European cuckoo, *Cuculus canorus,* the answer is no. In fact, when ectoparasites pass by accident into the plumage of the young cuckoo, they survive only a few hours. Young cuckoos are thus free of ectoparasites, contrary to all other young birds. *C. canorus* individuals do not acquire ectoparasites (which are specific to this species) until they mate in their second year. As Valkiunas and Iezhova (2000) emphasize, this absence of ectoparasites in young cuckoos is one more advantage that cuckoos themselves gain by being parasites.

Coevolution in Other Parasitic Associations

Even if certain aspects of the association with a host are novel, the use of the term *parasite* for cuckoos is not misplaced. There is an exploitation of a habitat (the nest constructed by the host), an expropriation of energy (food originally destined for the host young), a persistent interaction (the relationship that lasts for more than three weeks between the young cuckoo and its adoptive parents), and finally virulence (the loss of reproductive success by the parasitized couple). The cuckoo is therefore comparable to any other pathogenic agent, such as the human influenza virus or the sheep liver fluke, that uses living organisms simultaneously as a habitat and an energy source.

The arms race between cuckoos and their hosts can be used as a model to understand how all other parasitic associations work. In all cases, adaptations selected in the population of the pathogenic agent elicit counteradaptations in the host genome and vice versa.

Although in cuckoos and their hosts we observe only morphological and behavioral adaptations, most associations entail physiological adaptations, especially those related to immunity. When the association between a parasite and its host involves cellular contact between the two species, molecules of very distinct natures are in contact, and host cells have been specialized in the course of evolution to differentiate self from non-self.

I should note in passing that the distinction between self and non-self was already developed in the host species of cuckoos. But the distinction between "non-self" eggs (those of the cuckoo) and "self" eggs rests only on visual stimuli. In immunity, the membrane molecules of specialized cells identify foreign molecules. This identification triggers cascades of signals that lead to a defense in which either free molecules (antibodies, interferons) or cells (lymphocytes, macrophages) attack the intruder.

This simple observation of molecular involvement is the key to the arms race between pathogens and hosts. Selection in pathogens under pressure from immune defenses leads to three major classes of adaptations. (1) The pathogen attempts to be as imperceptible as possible within the host (a strategy completely analogous to egg mimicry by cuckoos), for example, by coating itself with molecules borrowed from the host. (2) The pathogen confronts the immune system by launching immunosuppressive or immunodepressive molecules against it. (3) The pathogen continually changes surface molecules in order to confound and outpace the host immune reactions.

One of the most original arms races is probably the one that entails programmed cell death (apoptosis), a genetically programmed "suicide" that cells can undertake. Apoptosis is well known to play a role during development of organisms, but we have also known for several years that apoptosis is employed to defend against parasites. If a cell is infected by a parasite and can cause its own death before the parasite has time to multiply, it will cause the parasite to die also and thereby protect the other cells of the organism. However, this adaptation has triggered selection for a counteradaptation by viruses and bacteria, which can send the cell molecular signals that block apoptosis (see James and Green 2004). Conversely, when it is favorable for parasites to cause apoptosis in host cells that are attacking them, they can do so. A specific protein produced by *Schistosoma mansoni* infective stages induces apoptosis in a particular population of host cells that is trying to kill them (Chen et al. 2002).

When pathogens respond to host immune defenses, they cannot avoid triggering a potentially perpetual arms race. Thanks to natural selection and to the genetic variation that permits it, the hosts are going to take countermeasures that will elicit counter-countermeasures ad infinitum. This coevolution can persist for a very long time, and it is probably the reason why there are so many pathogens and why there will probably always be so many. The most frequently asked question is, Will arms races end on their own when the measures deployed have achieved a certain balance or equilibrium? Zahavi and Zahavi (1997, 185) answer in the affirmative: "There are undoubtedly 'arms races,' but we think they quickly reach a stalemate. . . . Most host–parasite systems that exist today are in a state of equilibrium." This fact does not end biologists' discussions on this difficult subject, in which experimentation is impossible.

In principle, coevolution between parasites and hosts does not differ from that between predators and their prey: cats have excellent adaptations to catch mice, but, because the mice have excellent adaptations to avoid cats, both cats and mice continue to exist. However, in its details, host–parasite coevolution has many distinctive features.

The Red Queen Hypothesis

Lewis Carroll and Leigh Van Valen

Over the past two decades, the Red Queen has become a familiar figure in articles, books, and television shows about evolution.

The Red Queen is borrowed from Lewis Carroll's novel *Through the Looking Glass*. In this marvelous tale in which Alice is surrounded by all sorts of imaginary characters, there is a scene in which Alice takes the Red Queen (a chess queen) by the hand. Both run along a path, and suddenly Alice notices that the surrounding countryside is not changing. Taken aback by this astounding phenomenon, Alice queries her friend. The Red

Queen then replies, "Now, here, you see, it takes all the running you can do, to keep in the same place." This is why the surroundings seem not to change. In fact, the Red Queen is not really stuck in the same place any more than she is stuck in Lewis Carroll's novel since her reappearance in a 1973 essay by the American evolutionist Leigh Van Valen, who proposed a hypothesis on "what makes evolution proceed."

That the operation of evolution by natural selection is accepted today by nearly all scientists at least in its major features is one thing. That all questions about it are resolved is quite another. How natural selection of the sort that Darwin described has been able to take life from the appearance of the simplest nucleic acid molecule, which arose by chance in the famous "primordial soup," to the most complicated species, humans included, is not always easy to understand. The most difficult question is to identify the forces that have pushed natural selection ever further, giving the impression that it is always seeking to make more and better species.

It is not in doubt that changes on the surface of our planet, especially climatic changes and continental fragmentation and drift, have played an evolutionary role. But despite the occasional tendency of certain biologists to magnify this role beyond what is reasonable, it is surely not simply upheavals of the physical, inanimate earth that by themselves led to the inexorably increasing complexity of life.

This is why Van Valen suggested that the main motor of evolution for all species is the presence of other species with which they compete: any change in a particular species modifies the environment of the surrounding species and forces them to adapt. This adaptation itself constitutes a change in the environment of other species, and so forth. Thus, if two species are competing, each time one of them acquires any advantage whatever by natural selection, this advantage changes the environment of the other and generates selection on that other species for a compensating advantage. This suggestion is completely defensible if we consider that, in any environment, available resources are limited, "finite," so if one species improves its adaptation, this improvement can only be to the detriment of species exploiting the same environment. Except, of course, if selection produces compensating adaptations in the other species. In this case, the overall quality of the adaptations does not change, just as Alice's surroundings do not change. We can understand why Van Valen named his hypothesis the Red Queen.

To justify his hypothesis, Van Valen used an argument from paleontology. With a data base of many types of species (from foraminiferans through trilobites and ammonites to mammals), he observed that the probability of extinction of any species had no relation to how long it had already existed. In other words, the most recent species appeared no better adapted than the oldest. That is, from the fact that the extinction rate within any particular group of species remained constant through time, Van Valen deduced that the newest species were no better adapted to their environment than the oldest. This conclusion is very surprising, because each group has acquired many new adaptations through the course of its existence. As in Lewis Carroll's novel, we really have the impression of a trajectory that, even while endlessly producing ever more elaborate adaptive changes, nevertheless leaves the species in the same place insofar as the quality of their adaptations goes. Mark Ridley (1994, 594) wrote, "species do not evolve to become any better (or worse) at avoiding extinction."

The metaphor of Alice and the Red Queen has had wide resonance. As I noted earlier, it best matches biological reality when the number of partners engaged in an arms race is reduced. If we imagine a narrowly specific association of two species, any new adaptation of one exerts selective pressure on the other and forces it to adapt. In a host–parasite system, for instance, each time the parasite acquires a new "weapon" (such as a new antigen), the host is led to produce a new defense (a new antibody, directed at this antigen), and so forth. Although the Red Queen hypothesis applies to all biotic interactions, persistent ones, often characterized by few partners and great fidelity of these partners, are the ideal systems in which to study this hypothesis.

The Red Queen hypothesis spurs us to ask two questions, however: Is the landscape truly unchanging? Will the race stop some day?

Let us consider again the metaphor of Alice and the Red Queen in their moving landscape (which represents "other species" for Van Valen). Imagine that Alice and the Queen accelerate; at this moment, they seem to pass by the landscape. Imagine now that they slow down; the landscape will then seem to be passing them.[4]

If two or more species coevolve according to the Red Queen hypothesis, it is very likely that lags will arise among them. Such lags do occur and are called evolutionary lags (Stenseth and Maynard Smith 1984). "When the evolutionary lag grows, the species loses contact with its competitor or competitors, as does a lagging runner who sees the adversary inexorably pull away. In both cases, if the separation surpasses a certain value, it will never be closed again subsequently" (Combes 1999, 52).

The notion of evolutionary lag is especially important in persistent interactions. As I have shown (chap. 1), parasites and hosts do not have the same weapons. Parasites in general benefit from a higher mutation rate and shorter generation length than do their hosts. It is therefore highly likely that, in most such systems governed by a Red Queen process, at any given moment neither partner possesses the traits that would confer ideal adaptation. Natural selection can be very quick; other times it can be very slow, in relation to environmental changes, whether they depend on other species or on physical factors that can be almost instantaneous on an evolutionary time scale. So we should never assume that the state in which we observe an association is an equilibrial state, especially if the "partnership" is recent.

The words *indefinitely, continuously,* and *perpetual* figure prominently and repeatedly in Van Valen's article. Implicitly, he thus conceives of the Red Queen process as never-ending. This opinion is not universally shared by evolutionists. For instance, Zahavi and Zahavi (1997) believe that these arms races should rapidly arrive at an impasse (see the previous section). According to them, host–parasite systems as currently seen are in equilibrium. For example, observing that certain individuals of a particular bird species (Zahavi and Zahavi cite the reed warbler as a case in point) reject cuckoo eggs whereas other individuals do not, these authors do not see genetic differences (susceptible to selection) as causal, but rather the simple fact that the rejectors are at least two years old and have been able to memorize the appearance of their own eggs from their first reproductive period. In this case, obviously, the arms race can be considered ended.

The Consequences of the Red Queen Hypothesis

In the middle of the twentieth century, the great British evolutionist J. B. S. Haldane[5] pointed to the importance of genetic diversity in defense against pathogens. Half a century later, Steven Frank (2000, 169) wrote that "attack and defense invariably favor diversification." If one accepts the hypothesis that arms races exist and are at the heart of persistent interactions, diversity is an indispensable trait, if only to be able to parry the new "inventions" of an adversary. Robert Barbault (1994) stressed that, if a host population produces an accumulation of identical genotypes, the parasites will quickly adapt to them, and conversely on the part of the hosts if the parasite population has many identical genotypes. Genetic diversity therefore appears to be essential baggage if a species is to retain its rank in the merciless world of natural selection. What is uncertain is whether arms races work the same way in parasitism and mutualism. We usually assume that diversity is the best weapon for both partners in parasitic associations, but this need not be the case in mutualisms. Bergstrom and Lachmann (2003), for example, show by a mathematical model using game theory that, in mutualisms, "slow evolution may actually lead to favorable outcomes." This hypothesis has been called the Red King effect to contrast with the Red Queen theory, in which coevolutionary processes favor high evolutionary rates. However, the Red King effect needs empirical confirmation.

If we follow Haldane's and Frank's reasoning a bit further, we can predict that the mechanisms generating genetic diversity should be stronger the more intense the arms race is. For example, a host species should logically need greater genetic diversity if it finds itself subjected to very strong pressure by parasites, and vice versa.

Now, in higher species, thanks to the recombination of genes during meiosis, sexuality is a powerful mechanism for increasing genetic diversity originally spawned by mutations. We are then entitled to ask the question, Does sexuality exist in host species simply to help them struggle against parasites, and does it exist in parasite species only because it helps them resist host defenses?

Most evolutionists would say yes, although the question is still controversial. In fact, the "goal" in an arms race is not to select an ideal genetic combination that can fight an adversary but to renew genetic diversity continually. The important thing about recombination, which sexuality allows, is that it generates new host genotypes able to do combat against weapons, even unexpected ones, that have been selected in the adversary population. The production of diversity (by mutation but also by recombination) is therefore a process that opposes the homogenizing influence of natural selection. The importance of this mechanism for a species to survive in the long term is that it resists natural selection, even if this statement seems paradoxical. Any adaptation that is too tight renders its bearer fragile in the marketplace of evolution as well as in the workforce.

Controversy: The Evolution of Virulence

Does Virulence Increase or Decrease through Time?

It has long been recognized that a host–parasite system that has had little time to evolve is characterized by strong virulence (see chap. 3 for the definition of virulence); that is, it sharply reduces the reproductive success of the hosts. From this observation we have deduced that the trait "strong virulence" is primitive and that, with the passage of time, it is replaced by the evolved trait "weak virulence." Some authors go so far as to believe that associations of a mutualistic type are the normal end point of such evolution. Pathogens, in other words, should not kill the goose that lays the golden eggs.

In the 1980s and 1990s, new reflection on this issue profoundly enriched the concept of the evolution of virulence. Above all, it was recognized that natural selection must maximize the reproductive success of the parasite, and if the parasite easily finds new hosts to infect, weakening or even death of the individual parasite will not lower its reproductive success. There is therefore no objective reason why parasite virulence need always diminish with time (Anderson and May 1982). Virulence is an adaptive trait like any other.

It suffices to return again to the metaphor of filters of encounter and of compatibility in order to understand that virulence can either increase or decrease from generation to generation, depending on initial conditions:

1. When a host–parasite system arises (for example, after a migration or behavioral change of a potential new host), the filters can be slightly open, either because encounters are rare or because host immunity is initially effective. In this case, selective pressures operate mainly on the parasite population and not on that of the host. If parasite genetic diversity allows it, there will be selection of genes that will increase encounter frequency and/or evade the immune response, so virulence can only increase from generation to generation.

2. If, on the contrary, the filters are wide open, either because encounter frequency is high or because host immune defenses are ineffective, selective pressure will fall primarily on the host population rather than the parasite population. If host genetic diversity is high enough, there will be selection of traits that lead to either fewer encounters and/or a more effective immune response. Virulence can then only decrease through time.

In human diseases, the dominant impression is that emerging diseases are highly virulent and that this virulence tends to diminish through time. In support of this view, we should recall that there is a striking contrast between the gravity of parasitic diseases in regions where they are endemic and their impact when they attack populations that have never before faced them. This is why pathogens of all sorts have played a major role in human history.

After the Renaissance, the Europeans who embarked on the conquest of new lands were often aided by diseases they carried with them, but they were also sometimes stopped in their tracks by diseases they discovered in prospective colonies.

Conquerors Helped by Pathogens

It is believed that 50 million American Indians died from viruses and other pathogens brought by Europeans (Black 1992). Similarly murderous slaughters marked the conquest of Polynesia. Pathogens imported by Europeans killed 80 percent of the initial population of the Marquesas and 90 percent of that of the island of Rapa. In Polynesia, history has retained the names of ships such as the *Britania* and the *Magicienne,* whose passages were followed by catastrophic epidemics of influenza, dysentery, smallpox, and typhoid fever (Martin and Combes 1996).

Conquerors Hindered by Pathogens

Other pathogens, present in newly attacked host populations, have the opposite tendency—to oppose the new invaders. For example, in 1895 the French expedition to Madagascar lost 5,756 members; 25 of these soldiers died in combat, and the other 5,731 deaths were the result of malaria and other "local" diseases to which they were obviously less resistant than the natives. It is not an exaggeration to say that the fact that black Africa was barely explored and colonized before the first third of the nineteenth century was because, as soon as white invaders arrived, diseases such as malaria and African sleeping sickness ravaged them. These diseases also imposed a cost in illness and death on the natives, but this cost was lower than that paid by the immunodeficient Europeans (Aubry 1979).

Is Optimal Virulence Always Selected?

In many cases, optimal virulence (see chap. 3) is not selected, but the reasons vary. One reason is classically ecological: there cannot be adaptation to local conditions if gene flow from other populations constantly counteracts natural selection. To understand how such gene flow can influence adaptation to local environments, consider a nonparasitic example.

In the United States, the salamander *Ambystoma barbouri* is confronted with selection pressures that differ greatly depending on the habitat. When the salamanders are in permanent water bodies, they encounter carnivorous fishes. In temporary water bodies, fish are absent. The behavior of larval salamanders experimentally confronted with fishes differs depending on which of the two types of habitat the larvae came from (Storfer and Sih 1998). Larvae from permanent water flee or decrease activity in the presence of fishes, which gives them at least some protection. By contrast, larvae from fishless habitats do not display these antipredator behaviors. We can deduce from these observations that the antipredator behavior was selected only in those habitats with fish, which is entirely logical. However, another factor that affects the quality of the adaptation is how isolated the habitats are from one another. In other experiments, the researchers showed that salamander larvae from permanent streams near temporary streams had less effective flight behavior than those from permanent streams distant from temporary streams.

The explanation is simple. If salamanders move between neighboring sites, the resulting gene flow impedes the evolution of local adaptation. There is a continual dilution of "adapted" genes by "nonadapted" genes. It is possible to generalize from this salamander example. If a species exploits several distinct habitats and if populations exchange individuals between these habitats, it is impossible for optimal adaptation to arise in any one habitat; gene flow opposes selection.

When a parasite species exploits several host species, there is a mélange of genotypes in each generation. If, for example, humans and cattle are hosts of the same parasite species, optimal virulence cannot be selected in each host species. If one of the host species is more frequently exploited by the parasite, it is in this species that virulence will most closely approach the optimal level. In the "minority" host species, virulence can remain suboptimal or superoptimal. When infectious or parasitic diseases of humans are shared with other animal hosts, and if these hosts are dominant in the parasite life cycle, then virulence in humans can have little importance in parasite reproductive success. This fact explains why, with certain parasitic diseases common to humans and one or more other animals, the affliction can be more severe in humans than in the other species.

Another factor that opposes the evolution of optimal virulence is rapid variation of environmental conditions. The degree to which filters of encounter and compatibility are open is strongly influenced by the environment. For example, the success of the transmission of the liver fluke *Fasciola hepatica* depends heavily on weather conditions.

Fascioliasis is a water-transmitted disease of accumulation, whose gravity in cows and sheep (and in humans) increases with the number of infective stages ingested. In humid years, there is abundant reproduction in the intermediate snail host *Lymnaea trunculata,* and thus great dispersal of infective metacercariae in flooded pastures, and often prolonged survival of these metacercariae. The confluence of conditions favorable to the parasite opens the encounter filter to the point that the pathogenic effect is dramatically increased. Conversely, very dry years can also lead to great transmission frequency, because the two protagonists in the parasite life cycle, snails and livestock, then tend to concentrate in the few areas that remain moist.

The compatibility filter can also be affected by environmental conditions, because the quality of resistance by the hosts depends to a large extent on their general health and therefore on the availability of food. When animals (including humans) are malnourished, normally well-controlled parasites can quickly become highly virulent. For example, the semi-wild sheep of the tiny St. Kilda archipelago north of the Scottish mainland are periodically very sensitive to intestinal nematodes that are innocuous at all other times. The reason is that, when the sheep population surpasses a certain threshold, overpopulation leads to malnutrition. The malnutrition induces a lowering of immune defenses (the compatibility filter opens), and the proportion of parasites that manage to establish themselves increases substantially (Gulland 1992).

Sexual Selection
and Parasitism

7

Sexual Selection

The How of Sexual Selection

Modern ideas about biological evolution rest on a few simple principles. The first is that it is not individuals but populations that evolve. This means that traits acquired during an individual's life, such as calluses on the skin or hypertrophied muscles, are not passed on to descendants. Contrary to Lamarck, there is no inheritance of acquired characteristics. The second principle is that genetic information is not completely invariant from one generation to the next. Mutations create variation in populations: their individuals differ more or less from each other not only because of genetic recombination produced at meiosis but also because of new mutations. The third and last principle is that individuals undergo selective pressure. Confronted with the environment, some individuals have greater reproductive success than others. The result is that, over time, their genes spread in the population. Because new mutations arise routinely, the genetic structure of the population (the relative frequency of different genes) constantly changes. After many generations, individuals can differ noticeably from their ancestors in appearance, physiology, or behavior.

Evolution therefore results from two equally indispensable processes: genetic variation, which results from changes at the genotype level, and natural selection, which differentiates among phenotypes.

How could animals
choose resistant mates?

◊

WILLIAM HAMILTON *AND*
MARLENE ZUK (1982)

Selective pressures are the least controversial motor of evolution (among other possible motors, I have already mentioned global catastrophes). For example, if black mutants appear in a population of white moths, they will have an advantage if the environment (including the structures on which the moths rest) is dark. In fact, birds that eat moths detect white moths better in such a circumstance, because the white moths contrast more sharply than the dark ones with the environment. Conversely, the white moths have an advantage if the resting places are themselves light colored because it is then the black moths that are more readily seen and more frequently eaten. Beginning in the late nineteenth century, this type of selection pressure has been suggested as the explanation for the fact that the pepper moth became darker as industrial pollution darkened its resting places (such as tree trunks) in England. Some field experiments conducted in the mid- and late twentieth century to test the hypothesis were marred by errors and are contested (Hooper 2002). However, although the phenomenon of industrial melanism is doubtless more complicated than originally believed, none of the subsequent research undermines the basic reasoning about the role of differential predation by birds or the experimental demonstration that such predation occurs (Majerus 2003).

Sexual selection is a process in which differential success among several genomes depends on the choice of mating partners. This is a highly effective mode of selection. If, for example, at mating time females can choose among males or males can choose among females, and if a genetically determined trait (size, color, song) influences that choice, the frequency of that trait will increase or decrease in the population, depending on these choices. We can even imagine that an individual carrying some specific trait will never be accepted as a mating partner by the opposite sex. This is the ultimate strong selection against that trait.

Sexual selection probably exists in many species and not only in the most complex, highly organized ones, but it has been studied mainly in birds, fishes, and some mammals. The most frequently observed form is that of females choosing among males. In many birds, male mating candidates court females assiduously and more or less spectacularly, depending on the species. The females then choose one of these suitors, and it is obviously the genes of this individual that will be transmitted to the offspring.

This is the mechanism Darwin invoked to explain sexual dimorphism in some bird species. If the males and females are hard to tell apart in some species, by contrast in others the male is very brilliantly colored and the female is drab. Moreover, the males flaunt this ornamentation in elaborate displays accompanied by songs. If the rooster limits himself to displaying the feathers of his cappa magna, birds of paradise have much more spectacular behavior. Bruce Beehler (1990, 38) describes an encounter with males of the species *Paradisaea raggiana* courting a female: "How beautiful they are, with their yellow heads, bright green necks, and downy green breasts! They shake their long red feathers while parading before a female who came to the center of the group. . . . The movements of these birds, their colors, and their nuptial songs were then and still are today one of the most beautiful natural spectacles I have ever contemplated." After this drill, the males remain with lowered heads on tree branches, resembling, according to Beehler, a "resplendent fountain of colors." Then the female selects her mate, and the chosen male copulates with her. The birds of paradise that are so sexually dimorphic are polygamous: each male fertilizes as many females as he can and leaves to them the entire job of feeding the young.

Darwin's proposed explanation is that the choice of mates by the females constitutes a selective pressure leading to the brilliant male ornaments. It suffices, in fact, to assume that genetic variation for brilliant features is constantly renewed and that females tend to choose the most beautiful partners. In that case, genes for brilliant features will spread in the population, as will genes for showy strutting and beautiful singing. It is among polygamous birds that we expect sexual selection to be most intense, because in these species there is great variation among males in reproductive success, with certain individuals fertilizing many females and others none. Competition is therefore highly pronounced.

That beautiful males really are chosen by females has been demonstrated by many observations and also by an ingenious experiment. Anders Møller (1988) captured several dozen barn swallows (*Hirundo rustica*) in a Danish village and divided them into three groups. He shortened the tail feathers of individuals in the first group by 2 centimeters, lengthened the tail feathers in the second group by 2 centimeters, and left the lengths of the tail feathers of the third group unchanged. He performed these experiments by cutting the tail feathers at their midpoints, then sticking the pieces together with superglue. He cut and glued the tail feathers of the third group in order to control for the effect of the treatment independent of length. For each male, Møller then measured the elapsed time to acceptance by a female. The results were convincing: the males with shortened tail feathers took on average 12 days to copulate, whereas those with normal tails took 8 days and those with lengthened tails 3 days.

Of course, in cases in which the male chooses from among females, the mechanisms of sexual selection are identical but operate in the opposite direction.

In what follows, although I am going to talk essentially about sexual selection in hosts and the influence parasitism can have on this selection, it is worth noting that sexual selection can exist in the parasites themselves. For example, Sasal et al. (2000) revealed a probable case of sexual competition among males in a species of acanthocephalan worm that parasitizes marine fishes. The acanthocephalans are a small group of parasites with a heteroxenic life cycle (see fig. 3.1). In acanthocephalans, sexes are always separate and sexual dimorphism is pronounced, with females much larger than males. Each acanthocephalan male has two testicles. In general we would predict that investment in the size of testicles, which produce sperm, would increase when competition for mating with females increases. Sasal et al. showed that this prediction is borne out: after correcting the data for the sizes of both hosts and parasites, they found a significant positive correlation between the variables of testicular volume and percentage of males in the infrapopulation. The tighter the intermale competition, the greater the investment in testicular volume.

The Why of Sexual Selection

In ecological terms, to produce beautiful ornaments or to parade before females entails a cost. This cost is tied to the energetic expense (construction of the ornaments, execution of the parade) and to the risk of predation (predators can see the most striking individuals best).

In evolutionary terms, we are driven to ask the question, In birds, for example, what is the benefit of this brilliance, of these parades and songs? It does not suffice to say that the females prefer the most beautiful males, because the notion of beauty is, as far as we know, uniquely human. The choice behavior of the females can be explained only if they benefit from this behavior. The necessary condition for the females to benefit from their choice is that the brilliance constitutes a signal, that this signal indicates something, and that that something is useful in some way.

In order for brilliance to be a signal, it must vary in concert with some trait that would not be discernible without the signal. This varying actually occurs. It derives from the cost of the brilliance itself, because males cannot be brilliant unless they are in good health. Thus, brilliance or a successful parade signals a male in good health. Dull appearance or a lackadaisical parade, conversely, signals a male in poor health. Does this explanation suffice? No, because we must still discover what benefit the females gain from mating with males in good health.

As Møller, Christe, and Lux (1999) recall, there are three hypotheses. Two of them rest on parasitism.

The first hypothesis makes sense only in monogamous species, in which males help rear the young. In this case, by choosing a male in good health, a female has the highest probability that her partner will actively participate in feeding the young, and therefore in successfully rearing them, thereby ensuring that her own genes are propagated.

The second hypothesis can apply equally well to monogamous or polygamous species. It rests on the idea that the female can avoid infecting herself with parasites with direct life cycles. Many ectoparasites are transmitted by contact between individuals, so infection often occurs during mating. In birds, for example, if dull coloration or an uninspired parade signals that the male is infected, then by avoiding mating with him, the female preserves her own health and aids her reproductive success.

Finally, the third hypothesis is the most original. It is known as the Hamilton–Zuk hypothesis after the two scientists, William Hamilton and Marlene Zuk, who proposed it in 1982. This hypothesis suggests that, by mating with a brilliant male, a female is seeking to give good genes to her offspring, especially genes conferring resistance to illness. The choice is therefore not motivated by avoiding infection of the female herself during mating; rather, it is to ensure that the offspring of the female can resist pathogens.

In each of the three hypotheses, we see that the choice of a good partner has as its sole objective the transmission of the genes of the female to the next generation through the production of viable offspring. The females that have genes causing them to make good mate choices are at a selective advantage, because the choice increases the probability that their young will survive and transmit the "good genes" of the two parents. Of course, the "decisions" of the female must be understood as resulting from natural selection and not conscious reasoning, as my language has implied.

Parasitism and the Choice of a Partner

Good Females

In marine fishes of the group that includes pipefishes and seahorses, reproduction has some very curious aspects. When a male and a female are ready to reproduce, they engage in a sort of dance, tightly clasped, after which the female deposits a packet of ovules in a pouch that the male has on his abdomen. The male then fertilizes the ovules with his sperm and keeps the eggs in his pouch. Seahorse males even have a sort of placenta that nourishes the embryos, and it is only after the young have attained an advanced degree of development that the male "gives birth" to his offspring.

This is when a parasite intervenes—*Cryptocotyle,* a trematode. Species of *Cryptocotyle* have a complex life cycle. As it does for all trematodes, the life cycle of *Cryptocotyle* begins in a mollusc (for *Cryptocotyle,* it is a snail in the genus *Littorina*), in which it undergoes asexual multiplication that gives rise to the cercariae, which are small swimming larvae. The cercariae enter marine fishes and encyst there, becoming metacercariae. Finally, if the fish is eaten by a bird, such as a gull, the cysts open in the digestive tract and the metacercariae produce *Cryptocotyle* adults. Because of the obligatory presence of three successive hosts, no infection can occur between hosts at the same stage of the life cycle. For example, a fish cannot infect another fish.

It is, of course, during the second stage of the life cycle that the pipefishes play a role. They can be hosts to the metacercariae, and although each metacercaria individually is not strongly pathogenic, their accumulation (to as many as several hundred per individual) weakens the fish. An important detail is that the metacercariae are visible from the outside, appearing as black spots under the skin of the fish. It is possible to create false metacercariae by tattooing the fish with black ink.

G. Rosenqvist and K. Johansson (1995) studied this host–parasite system to try to determine if the number of parasite individuals affects the choice of a mating partner. They obtained two important results. First is that the presence of metacercariae significantly decreases the fecundity of pipefish females. Second is that the pipefish males prefer to mate with the least parasitized pipefish females.

This last observation has been repeatedly subjected to controlled experiments and validated. For example, false metacercariae have been tattooed on unparasitized females, which allowed the verification that it was indeed the presence of the parasites that determined choice by males, not the general condition of the female. Males avoided "falsely infected" females. It was also demonstrated that, although males avoid mating with these falsely infected females, they manifested no avoidance behavior toward other males, even other males covered with black spots. These experiments prove that the parasite affects selection of females by pipefish males and that the signal guiding this choice is visible parasite presence under the skin.

Why was this male behavior selected for? It is not to avoid their own infection, because the parasite is not contagious. Moreover, the fact that the males do not avoid infected (and falsely infected) males confirms that they are not avoiding their own infection. The explanation lies elsewhere. In terms of reproductive success, it is in the interest of a male to mate with the female who places the most eggs in his pouch, and, if parasites lower female fecundity, it is obviously in the interest of the male to choose an uninfected female. The advantage in choosing an unparasitized partner is thus immediate: she produces more offspring for the individual (male) who chooses her.

Good Males

Most birds harbor "lice" that in fact belong to the group Phthiraptera (which live on the body surface, as do children's lice).

Two species of bird lice found on the rock dove *Columba livia* (the common pigeon of cities worldwide) are common in feathers, of which they eat the most basal parts without damaging the skin or even apparently causing their hosts to itch. Infection by bird lice is direct; that is, individuals can transmit an individual parasite to a previously uninfected bird by simple contact, provided the contact is long enough. Sexual contacts during mating are prime moments for the passage of parasites from one host individual to another. Bird lice are therefore parasites that can be sexually transmitted, even if they can be transmitted outside of mating (for example, in the nest).

Dale Clayton (1990) studied whether these apparently inoffensive parasites can affect the choice of male pigeons by females. In fact, his studies have shown that bird lice are not nearly as innocuous as they seem. Their pathogenic effect derives from the fact that they eat the fine barbules located at the base of large feathers, which diminishes the insulating quality of the plumage. Infected pigeons cool more quickly and must eat more. Their poor thermoregulatory ability can even cause winter death if temperatures are especially low. Any bird that avoids infection has a better chance of surviving.

In the pigeon as in many other birds, males strut before females, and females choose their partners based on the quality of these parades. Clayton offered a female the choice of mating with either a louse-infested male or a louse-free one, and he repeated this experiment often with different males and different females. These experiments showed that parasitized males strut for a shorter period than uninfected males and that this is the signal that indicates the presence of bird lice to the

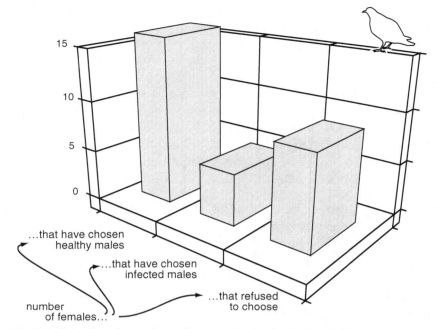

Fig. 7.1. Dale Clayton's experiment demonstrating that female pigeons favor as sexual partners males without parasites, even if the parasites (and their effects on plumage) are invisible to a human observer.

females, even if bird lice are visible only upon close inspection. In a statistically significant proportion of the tests, the female chose the male that strutted for the longest time (fig. 7.1). This work indicates that this behavior is indeed the result of selection to limit the risk of infection.

We see that, in this host–parasite system, sexual selection of a healthy partner confers on the individual making this choice an immediate advantage, that of not being infected by a parasite with a direct life cycle.

Parasitism and Longevity

In mammals, parasitism, sexual dimorphism, and longevity are often associated in a troubling manner. Moore and Wilson (2002) have clearly demonstrated this fact by comparative analyses conducted on a large sample of data drawn from the literature.

The relations between parasitism, sexual dimorphism, and longevity are as follows:

1. In most mammal species, males are more heavily parasitized than females.
2. The stronger the bias toward males in rates of parasitism, the more pronounced is sexual dimorphism (which measures the intensity of sexual selection).
3. The more pronounced sexual dimorphism is, the lower the longevity of males relative to that of females.

What can explain the association among these three traits? Two hypotheses can be proposed.

According to the first (fig. 7.2, hypothesis A), the decrease in male longevity could be the result of fights between males, which are one of the frequent components of sexual selection in mammals. According to the second (fig. 7.2, hypothesis B), sexual selection causes the increase in parasitism through the immunosuppressive effect of testosterone, and it is this increased parasitism that lowers longevity.

Moore and Wilson think that, "at least in part," the second hypothesis is correct. However, the reason for increased parasitism rates in males remains to be explained. Two hypotheses seem reasonable (fig. 7.2, hypotheses B1 and B2). Moore and Wilson believe that the immunosuppressive effect of testosterone is not the correct explanation for the increased parasitism of males. They suggest that, because males are larger owing to sexual dimorphism, they are bigger, easier targets for parasites. Moore and Wilson show that, in those rare mammal species in which females are typically larger than males (reversed sexual size dimorphism), the females are more heavily parasitized than the males. This fact favors the larger target hypothesis (B2) rather than the testosterone hypothesis (B1).

Moore and Wilson are aware that this line of reasoning collapses if sexual selection does not cause the other patterns (for example, if it is the consequence rather than the cause of increased male parasitism). We should recall that, in the hypothesis of Hamilton and Zuk (1982), sexual dimorphism is an adaptation allowing females to recognize "good males" (that is, those that resist parasites).

This discussion shows how difficult it is in biology to pass from correlation to causation. It also shows to what degree parasitism is nowadays at the heart of many questions about the evolution of life.

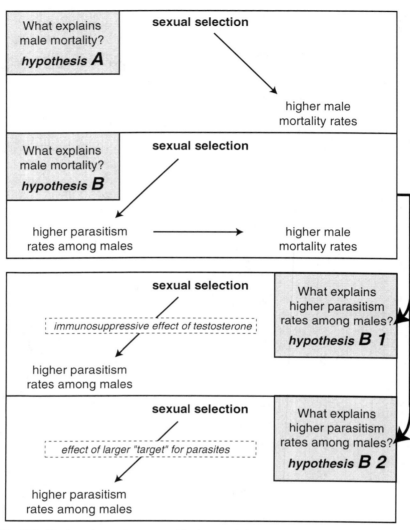

Fig. 7.2. The hypotheses that can explain the association, among mammals, between parasitism, sexual dimorphism, and longevity (see the text). (Inspired by data of Moore and Wilson 2002.)

Controversy: The "Good Genes" Hypothesis

The Logic of the Hypothesis

Suppose that, unlike the cases of the pipefishes and the pigeons, there is no immediate advantage to choosing an unparasitized partner. Would sexual selection then lead to such behavior?

As I said earlier, Hamilton and Zuk have hypothesized that this mate choice would ensure good genes for the offspring. In this case, there is no immediate advantage in choosing one partner over another. The advantage is deferred, or more precisely, it accrues to the descendants of the individual doing the choosing. As one may imagine, the Hamilton–Zuk hypothesis has been hotly debated. If the hypothesis is correct, we should be able to prove it by two main approaches. The first consists of studying a particular host–parasite model system. The second entails a comparative analysis among many species.

To validate the Hamilton–Zuk hypothesis by studying a particular system, it is necessary to satisfy four conditions enunciated by Andrew Read (1988): (1) resistance to parasites must be heritable (passed genetically from one generation to the next); (2) the parasite must negatively affect the color and/or parades of the males; (3) the females must choose less parasitized males; and (4) the reproductive success of parasitized males must thus be diminished. A study meeting these conditions was conducted on an American bird, the sage grouse (*Centrocercus urophasianus*), which forms leks.

The term *lek* designates, in birds, gatherings that arise during the mating season and during which the females choose males. In pheasants, grouse, and their relatives, for example, lek formation has as a result that only a small number of cocks actually do the mating for the whole population, while the other cocks do not participate at all in reproduction.

The pheasant/grouse case is interesting because males in these species do not aid in raising the young. Therefore, by choosing males that appear to be in good health by virtue of their brilliant plumage, the females are not gaining any help in raising their offspring. However, research by Gibson and Bradbury (1986) on sage grouse clearly shows that females choose males that make the most successful parades and that regularly join leks. We can therefore imagine that the females are either seeking to avoid infection by parasites that could be transmitted during mating or are seeking to give their offspring genes that confer resistance to such parasites.

Linda Johnson and Mark Boyce (1991) sought to verify the Hamilton–Zuk hypothesis by determining if there is a positive correlation between choices made by females and a parasite that is not directly transmitted, *Plasmodium pediocettii*, which causes malaria in pheasants and their relatives. The infection occurs by mosquito bite. Therefore, if an uninfected female mates with a parasitized male, he cannot contaminate her. As in any malarial infection, the malady in these birds induces a periodic feverish weakness coinciding with the time when the infected red blood cells burst to liberate parasites into the bloodstream. In this particular species of *Plasmodium*, this bursting and the associated weakness occur in the morning and only on certain days.

The study was conducted in Wyoming, where the sage grouse is abundant. To begin, the scientists captured grouse, assessed their physical condition, took a blood sample, tagged each individual so it could be distinguished with binoculars, then released all the animals. The leks, which form early each morning, were then followed throughout the mating season. For each male, it was therefore possible to determine how consistently he joined a lek, how often he paraded, and, of course, the number of times he mated.

The study showed that the main factor affected by the parasite is participation in the lek. Males infected by *Plasmodium* do not join leks on certain days. These days probably correspond to the periodic bursting of infected red blood cells. Moreover, the study also proved that, in making their choices, females very heavily weigh the frequency of participation in leks. The conclusion is obvious, and the logical sequence of cause and effect is as follows: (1) the presence of the parasite in males lowers their participation in leks; (2) this lowered participation reduces their attractiveness to females; and (3) therefore parasitized males mate less often and so do not transmit their genes as frequently.

It is interesting that, when they do participate in leks, the parasitized males do not appear to parade less actively than the uninfected males. This observation is easily explained by the fact that, on those days, the *Plasmodium* is "silent" in the medical sense of the term. The signal the females use to make their choices is therefore frequency of participation in leks, not frequency and vigor of parading when participating.

Figure 7.3 shows that the number of matings differs greatly between infected and uninfected males, over an observation period longer than a month. It also shows a certain time lag: parasitized males rarely have access to females until nearly two weeks after the first matings by uninfected males. The researchers believe that parasitized males are therefore mating either with very young females or with females with more or less compromised health. In either case, such females would rear smaller clutches than older, healthy females. There is thus an additional transmission disadvantage for the genes of males infected by *Plasmodium*.

Are Read's four conditions satisfied?

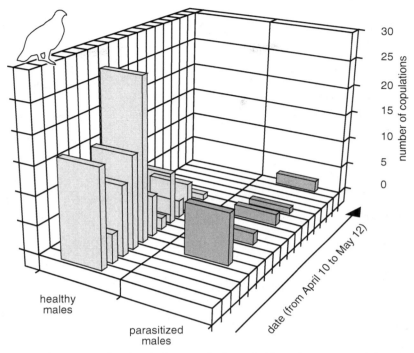

Fig. 7.3. Distribution of matings in the reproductive period by male sage grouse infected with a *Plasmodium*, and by uninfected males.

1. With respect to the heritability of resistance, it is possible to extrapolate from data obtained in other host–parasite systems. In humans, for example, several studies have shown genetic diversity with respect to resistance to malaria: the gravity of infection by *Plasmodium falciparum* can differ by a factor of 10 among individuals, depending on which genes they possess. It is therefore reasonable to believe that birds have a similar genetic basis for malaria resistance.

2. As for the effect of the parasite on the quality of the parades, the study by Johnson and Boyce is conclusive that the frequency of lek attendance is inversely related to degree of infection.

3. With respect to female choice, the results of the research are especially convincing.

4. Finally, for the reproductive success of males, this has not been directly evaluated by genetic analysis, but it is easily deduced from the two previous observations.

It is thus clear that the research of Johnson and Boyce supports the Hamilton–Zuk hypothesis.

If the Hamilton–Zuk hypothesis is correct, a logical prediction is that sexual dimorphism should be selected for when the benefit is very great, that is, if the species is attacked by many pathogens. In other words, there should be a positive correlation between the male degree of brilliance and the overall level of parasitism in the host species under consideration. When parasite pressure is typically weak, sexual dimorphism should be weak, and where parasite pressure is strong, sexual dimorphism should be pronounced.

Comparative analyses examining the variables of parasite richness and male brilliance have given inconsistent results. Some studies tend to support the Hamilton–Zuk hypothesis, others to contradict it.

Problems with the Hypothesis

I would be remiss not to say that there are several problems with the Hamilton–Zuk hypothesis. I do not mean by "problem" the fact that there are exceptions, nor that factors other than parasitism can produce sexual dimorphism. For example, the fact that certain parrot species have females as strikingly colored as males should not be considered as an argument against the hypothesis, because such a result could arise from all sorts of evolutionary pressures having nothing to do with parasitism. Similarly, the fact that certain comparative analyses, especially of fishes, have not yielded a relationship between dimorphism and parasitism is not surprising. The real problems rest on theoretical considerations. I will cite two of them.

The comparative analyses tending to confirm the Hamilton–Zuk hypothesis are based on the number of species of parasites. Poulin and Vickery (1993) pointed out that, for the Hamilton–Zuk hypothesis to be supported, the parasites would also have to be common, which is obviously not always the case. If the parasites are rare, females will be unable to distinguish resistant males from susceptible males that happen not to be infected, so sexual selection will not operate. Moreover, we know that parasites are almost always distributed in aggregated fashion among host individuals. This distribution can mean that heavily infected males are inferior and will be excluded from reproduction even before females have a chance to make their choices. In this case, it is not sexual selection that would cause the transmission of good genes but the pathogenic effect of the parasite.

The second theoretical problem arises from what Folstad and Karter (1992) call the handicap of immunocompetence. What is this? At least in vertebrates, the traits that constitute the signal for females—that is, the brilliant colors or vigorous courtship behavior—are secondary sexual characters, and as such their expression rests on the sex hormones, especially testosterone. Even weapons like horns or antlers, which males of some species use to fight one another over possession of females, are testosterone-dependent traits. Now, testosterone has immunosuppressive properties, which logically should lead to the facilitation of infection by parasites.

In fact, research conducted on wild animals, on domestic animals, and in human populations shows that, in many instances, males are more affected than females by parasites. It has been demonstrated experimentally that there can exist a positive correlation between the host concentration of testosterone and parasite success. Elevated concentrations of androgenic hormones therefore confer on individual males the advantage of weapons or attractive traits for competition but simultaneously cause the disadvantage of increased vulnerability to parasites.

We find ourselves facing the following paradox: sexual selection favors reproduction by individuals supposedly most resistant to parasites, but at the same time, the involvement of testosterone in the signaling mechanism renders these individuals more vulnerable to the same parasites.

Another paradox, called the car–house paradox, may give the solution. In human society, any given individual has, in principle, finite resources. Therefore, if he or she invests in the purchase of a beautiful house, fewer resources are available for the purchase of a luxurious car, and vice versa. In fact, we observe that it is almost always the same individuals who have both expensive houses and expensive cars, simply because their wealth is so great that they can sustain both expenses. Transposed to the domain of host–parasite relationships, the car–house paradox becomes the "resistance–testosterone" paradox in the sense that genetically superior individuals can simultaneously invest in resistance to parasites and in testosterone-dependent sexual traits. As Westneat, Hasselquist, and Wingfield (2003, 322) write, there is "a growing literature on the relationship between sexual signals, testosterone, and immune response." This fact clearly indicates that the question is complex, as is anything that touches on immunity. For example, the research of Hasselquist et al. (1999) and Westneat, Hasselquist, and Wingfield (2003) on a Kentucky population of red-winged blackbirds (*Agelaius phoeniceus*) casts doubt on the role of testosterone. In this research, testosterone does not seem to inhibit the immune response, and immunocompetence does not appear to be associated with male sexual traits such as territory defense during the mating season.

The three processes that I have baptized "quest for good males," "quest for good females," and "quest for good genes" are not necessarily mutually exclusive. In particular, the quest for good genes can underlie all cases. If pigeons have defense mechanisms against bird lice and if these mechanisms are heritable, females that avoid louse-ridden males can also confer on their young the corresponding resistance genes. Similarly, pipefishes that prefer to mate with the least parasitized females can also add to the immediate advantage of a large brood of offspring the additional advantage of giving them resistance genes.

We can conclude that today there is no doubt that parasitism influences sexual selection, when the choice of an uninfected partner gives an immediate advantage to the individual making that choice. By contrast, with respect to the quest for good genes for offspring, intraspecific studies of particular species and interspecific comparative analyses give evidence both for and against the proposition. Møller, Dufva, and Erritzoe (1998) have modified the data of comparative analyses by using as a variable not the number of parasite species but the developmental degree of immune system organs, such as the spleen and, in birds, the bursa of Fabricius. This change effectively removes from the analysis benign parasite species that do not exert selective pressure on their hosts and therefore cannot be involved in sexual selection. By thus retaining in the analysis only the size of the organs involved in resistance, the authors conclude that resistance to pathogens is indeed a real determinant of sexual selection.

Parasites in Space and Time

The Space of Hosts

No parasite species is everywhere. Neither is any free-living species. There are constraints that determine that a particular parasite uses a particular host and not other ones, that a particular parasite is rare while another is common. For a parasite, the result of these constraints is expressed by four parameters: abundance, distribution, specificity, and richness.

Abundance

Among parasites as among free-living organisms, some species are common and others rare. In either case, one of the major goals of ecology is to discover which factors cause abundance or rarity.[1] Figure 8.1 depicts an example: among flatworm species that parasitize the common frog *Rana temporaria* (here studied in the Pyrenees), there are very abundant species and others much rarer. (It suffices to compare the digestive tract trematode *Opisthioglyphe rastellus* with the lung trematode *Haematoloechus pyrenaicus*.)

Comparative analyses aid identification of causal factors of abundance or rarity. For example, when we compare the many species of parasites, the most frequently recognized correlate of abundance/rarity is body size (the larger the size, on average, the lower the abundance). This result conforms to similar findings for free-living organisms such that body size "explains" 90 percent of the variation in abundance. When we compare among hosts, the best-

One of the main goals of ecologists is to explain the abundance and distribution of organisms over space and time.

◊

ROBERT POULIN (1998)

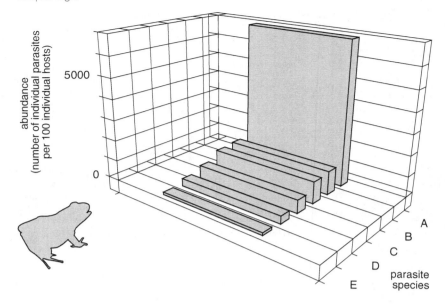

Fig. 8.1. Very different abundances of five parasites of the common frog in the Pyrenees, determined from examination of slightly more than 1,000 specimens. (A) *Opisthioglyphe rastellus;* (B) *Gorgodera euzeti;* (C) *Gorgoderina vitelliloba;* (D) *Polystoma integerrimum;* (E) *Haematoloechus pyrenaicus.*

known correlate of parasite abundance is host density: the denser the host population, the greater parasite abundance. As Arneberg, Skorping, and Read (1997) note, the former result suggests the existence of regulation of number of parasite individuals within individual hosts, whereas the latter suggests that parasite transmission is facilitated when hosts are abundant.

Of course, factors responsible for abundance or rarity differ among taxa (groups of related species) of hosts and among taxa of parasites. They are far from adequately understood. Despite the existence of general "rules," there seem to be unexplained disparities among abundances of different species. Progress in this domain is essential to the control of parasitic diseases of humans and domestic animals.

Distribution

Whether parasites are common or rare, a second question arises with respect to their relationship to space. This is the matter of distribution.

We can use the metaphor of a pool table with a series of holes and a bag of billiard balls that can be emptied over the table (fig. 8.2). When all balls have fallen into holes, we can analyze how they are distributed. In general, three situations are possible. (1) The balls are distributed as evenly as possible; for example, if there are 100 balls and 100 holes, there would be one ball in each hole. (2) The balls are distributed according to a Poisson distribution (that is, randomly), if there is no force causing any particular attraction to or repulsion from any hole for any ball. (3) The balls are concentrated in certain holes (for example, all 100 balls in one hole); we call such a distribution *aggregated*. There are therefore three fundamental types of distribution: regular, Poisson (random), and aggregated.

Questions about parasite aggregation arise at several levels. At the host species level, how are a group of parasites distributed spatially among available host species? At the level of individual hosts, how are a group of individuals of a given parasite species distributed spatially among host individuals? At the microhabitat level, how are a group of individual

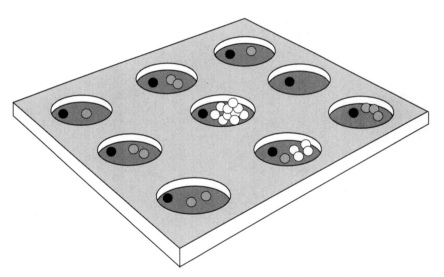

Fig. 8.2. The parasite pool table. The black balls have a regular distribution, the gray balls have a random (Poisson) distribution, and the white balls have an aggregated distribution. This diagram can be understood at three scales: the holes can be host species (in which case the balls are parasite species), host individuals (in which case the balls are parasite individuals), or microhabitats within a host individual (in which case the balls are also parasite individuals).

194 parasites within one host individual distributed spatially within that host individual?[2]

Consider a parasite taxon and a host taxon. How are the parasites distributed among the hosts?

Antoine Pariselle (1997) studied fishes in the family Cichlidae in western Africa and certain of their parasites, monogeneans of the family Ancyrocephalidae (fig. 8.3). The results are of great interest with respect to many aspects of parasitism, from species richness (number of species) and host specificity to evolutionary host–parasite relationships. Here I cite only a small part of this research dealing with 26 fish species and 74 monogenean species. Figure 8.3 shows the distribution of ancyrocephalid monogeneans among 5 host fish species (I am showing only those for which the sample reached 200 individuals and 10 sites). These data show that the species are distributed in highly aggregated fashion; for example, host species A harbors between three and four times the number of monogenean species as host species E. This result becomes even more striking when we consider just the monogenean genus *Cichlidogyrus* (darkened sections of bars in the figure): we see that, in fish species A, all monogeneans belong to this one genus.[3]

Now consider not a parasite taxon and a host taxon but a population of individual parasites (that is, all of the same species) and a population of host individuals. How are the balls distributed among the holes (this time, the balls and holes represent individuals)?

H. D. Crofton was the first to show, in 1971, that individual parasites are almost always distributed in aggregated fashion among host individuals.

Figure 8.4 depicts an example: the remarkable distribution in Australia of a population of the trematode *Simhatrema simhai* in the sea snake *Disteira kingii*. Of a total of 496 parasite individuals collected from 30 individual snakes, more than 430 were found in one individual (Holmes 1989). The *Simhatrema* example is an extreme one, but parasitological studies, whether of wild or domestic animals or of humans, generally reveal parasite distributions that are more or less aggregated.

Individual parasites are aggregated for three basic reasons: (1) the distribution of infective stages may be aggregated, generating aggregated infections among hosts; (2) parasites multiply after infection; or (3) the host population is heterogeneous with respect to susceptibility to parasites.

Reason 1 is easy to understand. If the infective stages are not dispersed in random fashion in the environment, each infection episode can involve several parasites, thus generating an aggregated distribution. The life cycles of the trematodes *Leucochloridium paradoxum* and *Microphallus pygmaeus* provide examples of this type of "lottery" infection. In most trematode life cycles (see the preceding chapter) there is multiplication in an initial mollusc host, which produces cercariae that disperse in the external environment and enter a second host; the second host species varies among trematode species. The second host is then eaten by a third host (the "definitive" host), in which the parasite becomes sexual and reproduces. In *L. paradoxum* and *M. pygmaeus*, the cercariae, produced by the hundreds, do not leave the mollusc and instead become infective in it. When a bird eats an infected mollusc,[4] it hits the lottery jackpot, so to speak, because it acquires hundreds of individual parasites (genetically identical, if they all arose from a single miracidium).

Reason 2 is also easily understood. If the parasite multiplies within the host, then only a single infective-stage individual can generate a substantial population in its host individual. With the malaria *Plasmodium*, probably only a lone sporozoite (the stage that, when injected by a mosquito, infects humans) suffices to give rise to billions of descendants. Although microparasites (viruses, bacteria, protists) are most likely to undergo multiplication after infection, the phenomenon is also observed in more complex parasite species. The multiplication of trematodes in their first mollusc host is an example. A much rarer sort of in situ reproduction is the "internal" sexual reproduction of monogeneans. For example, we find it in the species *Eupolystoma alluaudi,* parasitic on a small toad of the Sahel region of Africa. In *E. alluaudi,* one or two initial individuals suffice to populate the bladder with several hundred parasites in a few months, simply because their eggs can hatch where they are laid (that is, in the bladder).

It is reason 3, however, that is the most interesting, because it applies to all parasites, including those that affect humans. In fact, every host population is highly heterogeneous because of variation among individuals in the degree to which the filters of encounter and compatibility are open. I discuss this aspect of parasite distribution in the controversy that ends this chapter, using as an example parasites of humans.

Aside from host species and host individuals, there is a final scale of analysis of distributions, that of microhabitats. In fact, within a single individual host, differential expression of the same genetic information gives rise to different tissues and organs. The distribution of parasites of the same species is very sensitive to these differences of expression. Much research has been conducted on the distribution of parasites in the intestine and also in fish gills (for example, see Combes 2001). Not only are parasites localized, depending on the species, in the intestine or the gill, but they are further specialized to show strong preferences for certain parts of

Fig. 8.3. Distribution of monogeneans of the family Ancyrocephalidae (dark bars plus light bars) and of the genus *Cichlidogyrus* alone (dark bars only) in freshwater fishes of western Africa (from the data of Pariselle 1997). (A) *Tilapia guineensis;* (B) *Tilapia zillii;* (C) *Hemichromis fasciatus;* (D) *Oreochromis niloticus;* (E) *Sarotherodon melanotheron.* The numbers in parentheses indicate the number of individual fishes examined; the numbers in brackets indicate the number of different locations sampled.

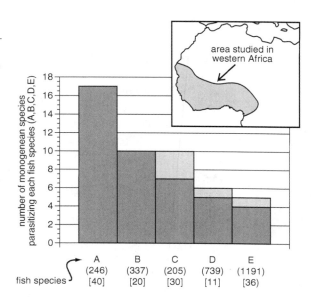

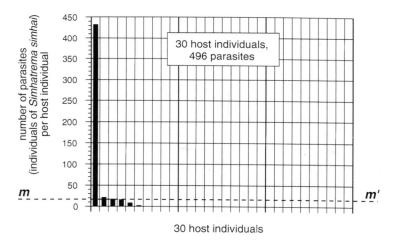

Fig. 8.4. Aggregation of parasites in host individuals: example of a trematode parasite of sea snakes. The line m–m' (mean number of parasites per fish calculated from a sample of 30 individuals) shows the distribution that would have resulted if each fish had received the same number of individual parasites.

these organs. These preferences, often observed to be remarkably fine in parasites, are explained by the heterogeneity of resources (substrate, nutrients, etc.) and the environment (currents, pH, etc.), which determine the morphology and physiology of the hosts.

To end this discussion of distribution, I can say that, if we were able, analogously to Laplace's demon, to "see" and locate every single living parasite in the biosphere at some specific instant, and if we were able to determine at that instant how they were distributed among host species, among host individuals, and among microhabitats within host individuals, we would observe that the distributions at all scales are aggregated. In an ecosystem, certain free-living species harbor many parasite species; others harbor few. In a host population, certain individuals are heavily parasitized; others are not. In an individual, some organs are heavily invaded by parasites; others are quite neglected.

Specificity
The Malaysian Swiftlets

If we return to the metaphor of the pool table onto which we throw billiard balls, parasite specificity is represented by the number of holes that contain the same type of ball (that is, the same parasite species)—this is the *host spectrum*. Parasite richness is exactly the inverse notion, that is, it is seen from the host perspective: the number of different kinds of balls (parasite species) found in the same hole (fig. 8.5).

When we speak of parasitism, the concept of narrow parasitic specificity (small host spectrum) is readily advanced. However, it is important to place this concept in a more general ecological framework. In fact, specialization to exploit a particular environment characterizes all species, not just parasites. No species, be it bacterium, mollusc, or bird, is found in every habitat. Perhaps humans are the most ubiquitous of all. But even they are incapable of living everywhere—for example, they cannot inhabit the sea floor or hot springs.

Parasitic species do not differ in this regard. On the one hand, they are specialized. On the other hand, they are specialists to different degrees. What actually constitutes a host spectrum?

Imagine an ecosystem composed of five free-living species (A, B, C, D, and E in fig. 8.6) and one parasite species, and let this parasite confront the filters of encounter and compatibility. Suppose the parasite cannot contact two of the five free-living species (A and E in fig. 8.6) for ecological or behavioral reasons. The only remaining possible hosts are species B, C, and D. Next suppose that, for metabolic or immunological reasons, the parasite

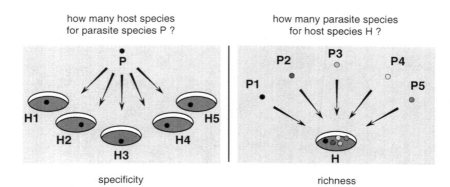

how many host species
for parasite species P ?

how many parasite species
for host species H ?

P

H1   H5
H2   H4
H3

P2   P3   P4
P1   P5

H

specificity

richness

Fig. 8.5. Specificity (*left*) and richness (*right*). Host species are represented by holes in an imaginary pool table; balls represent parasite species.

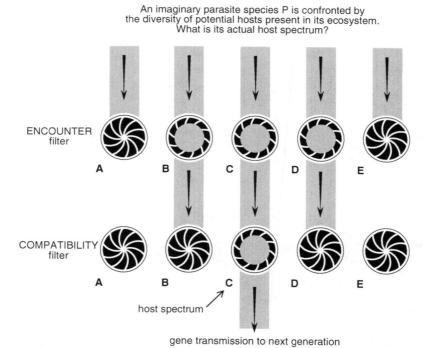

An imaginary parasite species P is confronted by
the diversity of potential hosts present in its ecosystem.
What is its actual host spectrum?

ENCOUNTER
filter

A   B   C   D   E

COMPATIBILITY
filter

A   B   C   D   E

host spectrum

gene transmission to next generation

Fig. 8.6. Constitution of a host spectrum drawn from species theoretically available in an ecosystem. The first level of selection is produced by the encounter filter, the second by the compatibility filter. The figure assumes that the final host spectrum is reduced to a single species (C), which is, of course, not always the case.

cannot develop in species B and D. The only possible host left is species C (of course, the host spectrum is not always reduced to a single species as in this example).

Figure 8.7 aims to sophisticate the concept of host spectrum, which is often represented by a simple species list. In reality, although there are cases in which a parasite develops in one host species and never in another (fig. 8.7, panel 3), there are also cases in which two host species (or several) play quantitatively different roles in the parasite life cycle (fig. 8.7, panel 2). Parasitism of some hosts can be solely accidental. All species in the host spectrum are not equally important in transmitting the parasite. In fact, if the host spectrum reflects a specialization for habitat at the level of host

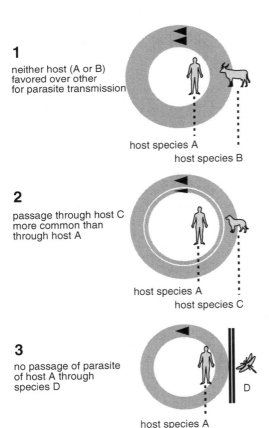

1

neither host (A or B) favored over other for parasite transmission

host species A

host species B

2

passage through host C more common than through host A

host species A

host species C

3

no passage of parasite of host A through species D

host species A

Fig. 8.7. The host spectrum of a parasite is not a list of hosts of equal importance for transmission. All intermediates (2) exist between two host species that the parasite uses indiscriminately, as if they were a single species (1), through total exclusion from one host (3).

species, specialization often goes much further, as is illustrated by research on parasites of swifts.

I now focus on a concrete example of habitat specialization (Tompkins and Clayton 1999) that shows that a living habitat offers an extraordinary diversity of environments, simultaneously at the levels of species, individuals, and microhabitats. This example will help clarify questions that arise about parasite specificity. It also has an advantage over many other studies in that it includes an experimental component.

The researchers chose to study lice frequently found in bird plumage. These parasites are typically very specialized; a particular louse species usually parasitizes just one bird species. Transmission is by contact, mainly between parents and nestlings, so we can ask if the specificity does not simply arise from the fact that contacts between individual birds of different species are rare. This idea can be tested by collecting parasites on one host species, then transferring them to a different host species. Controls (transfers between individual hosts of the same species) allowed the researchers to ascertain that it was not the transfer manipulation itself that caused subsequent failures of the transferred parasites.

Researchers selected as hosts for this study four species of cave swiftlets, insectivorous birds of Southeast Asia with the distinctive behavior that they nest in caves through which they navigate like bats, echo-locating obstacles. In the study area (Malaysia), cave swiftlets are parasitized by a total of six species of lice in the genus *Dennyus*. The swiftlets nest in groups of several thousand individuals in different types of sites, which prevents—or at least limits—contacts between individuals of different species. Birds can exchange parasites only when they collide in flight or when the parasites themselves move from nests of one species to nests of another that are not too far away.

Examination of 1,381 birds showed an apparently complex pattern of specificity, but in certain cases it was evident that a particular *Dennyus* species was never found in nests of certain swiftlet species. For example, *D. somadikartavi*, which parasitized 99 percent of the 240 individual *Collocalia esculenta* examined, was never found in the 398 individual *Aerodramus salanganus* examined. This case is a good example of a host barrier not breached in nature.

Tompkins and Clayton then conducted various transfers from normal hosts to novel hosts, as well as control transfers from normal hosts to normal hosts. Results were initially disconcerting: in certain transfers to novel hosts, survival of parasites was close to that of the control transfers. In other transfers to novel hosts, survival was very low. The explanation was found in a comparison of the diameter of the barbules of feathers in the different swiftlet species. Each time the barbules of the novel host were similar in diameter to those of the normal host, survival was high. When the diameters differed substantially, survival was poor. The most interesting observation was that, in certain instances, the "right" diameter existed on barbules of a novel host, but in feathers located in different regions than on the normal host. In these cases, survival was high, but the parasites changed microhabitat to use the feathers that provided them with barbules of the "right" diameter.

Figure 8.8 depicts results of transfers divided into two groups. Successful transfers (A) are those in which the recipient host species has barbules that do not differ in diameter from those of the donor bird by more than 2 micrometers. Generally unsuccessful transfers (B) were those in which the barbule diameter difference exceeded 2 micrometers.

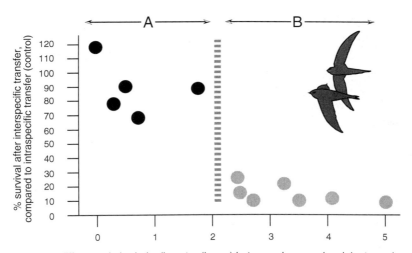

Fig. 8.8. Relationship between the difference of barbule diameters of donor species vs. recipient species and parasite survival in experimental transfers of *Dennyus* (after Tompkins and Clayton 1999, modified). (A) In the experiments marked by black points, the transfer between birds of different species succeeds in proportions similar to transfers between birds of the same species; one experiment (*leftmost high point*) even yielded higher survival rates for interspecific transfers. (B) In experiments marked by gray points, transfers between different bird species succeeded much less frequently than did transfers between birds of the same species. It is evident that interspecific transfers succeed only if the barbule diameters of the two host species are similar (difference is less than 2 micrometers [μm]).

The value of 2 micrometers thus constitutes a threshold. We can reasonably deduce from this fact that these parasites cannot resist wing beating and perhaps cleaning activities by the swiftlets unless the feathers have very nearly the perfect diameter for them to hang on. The adaptations of lice that allow them to grasp the plumage of their hosts are such that the margin of maneuver is extremely low when the morphology of available feathers is not just right for these adaptations.

Several conclusions follow from this study. The first responds to the objective stated by the researchers: it is not the impossibility of transfer that explains the narrow host spectra of these lice but the quality of the adaptation once they are transferred. The host spectra of species of *Dennyus* are therefore probably much more determined by the compatibility filter than by the encounter filter.

The second conclusion is that the parasites that were studied do not know bird systematics. The only thing they know is the resource these birds offer them, whatever the birds are named. Here the limiting resource is barbules of the right diameter. Compatibility is thus not always a matter of immunity; it can be related to just the habitat.

The authors also drew a third conclusion. Analyzing the bird phylogeny, they showed that there is no correlation between phylogenetic proximity of the hosts and parasite survival. It is therefore the present status of the resource and not the past heritage that determines specificity. The present status of the resource can derive from various sources, from its accessibility to its actual characteristics.

The example of the Malaysian swiftlets shows how natural selection has operated: in the relationship between *Dennyus* and the swiftlets, specificity depends on the diameter of the barbules. The importance of the feather resource was recognized early by Theresa Clay (1957, 132), who wrote that "a bird might become more easily infested with lice from another unrelated bird if both had similar feather structure."

Of course, we can ask the question, Why has natural selection not endowed the louse species with more versatile means of grasping feathers, which would allow them to colonize many more host species? We imagine it is a question of the balance between cost and benefit, but I will discuss this question in more detail, because the advantages and disadvantages of specificity are the subject of controversy that ends this chapter.

Specialists and Generalists

As the Malaysian swiftlet example shows, compatibility can be a matter of resources as well as escape from an immune system. Parasite specificity is but one form of the resource specialization that is a common trait of all species. Every species on this planet exploits the environment within certain limits. Giraffes do not inhabit ice floes, and cats do not eat slugs. The range of resources exploited varies greatly among species. Everyone thinks of the panda, which eats only bamboo leaves, or the koala that eats only eucalypt leaves. On the other hand, pigs and humans eat almost everything.

What is intriguing about parasites is that, in general, their specialization is narrower than that of most predators.

Figure 8.9 shows the host spectra of monogenean parasites of Mediterranean fishes.[5] In the figure, the value 1 for host spectrum means that this monogenean species exploits only one fish host species, the value 2 that it exploits two different fish host species, and so forth. We see that the great majority of monogeneans (134 species of 181 surveyed) use just one host species. None is found on more than 6 hosts. With respect to these figures, I should also add that detailed biochemical analytic methods have shown that, when a parasite appears to exploit several host species, it is often the case that what was thought to be one parasite species is actually a complex of "cryptic" species that are so similar morphologically that we cannot distinguish them but that are perfectly well isolated reproductively

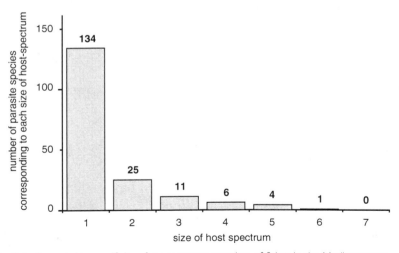

Fig. 8.9. Remarkable specificity of monogenean parasites of fishes in the Mediterranean (after Caro, Combes, and Euzet 1997, modified): 134 of 181 species are each found in only a single fish species.

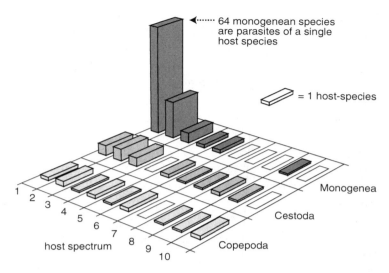

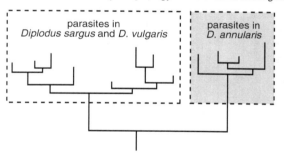

Fig. 8.10. Host spectrum of metazoan parasites of Canadian freshwater fishes. Sixty-four monogenean species of a total of 98 have been found in only one host species each. Observe that the part of the histogram depicting the monogeneans closely resembles the histogram depicting the marine monogeneans of the Mediterranean in fig. 8.9. (Data from Poulin 1992b.)

trematodes identified by morphology as *Macvicaria crassigula*

trematodes identified by morphology as *Monorchis parvus*

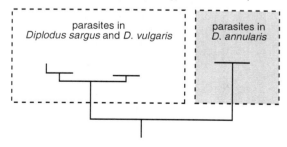

Fig. 8.11. Separation of "cryptic" species by phylogenetic trees based on DNA analysis. (Data from Jousson, Bartoli, and Pawlowski 2000.)

from each other. It is thus possible, for example, that if we conducted such biochemical research on monogeneans that are listed as parasitizing 5 or 6 different fish species, we would find we actually have 5 or 6 distinct parasite species. If this were so, we see that real host spectra would be even more restricted than they appear in figure 8.9.

Not all parasite groups are as highly specialized as the monogeneans, but host spectra of parasites belonging to other taxa are generally not very broad. Robert Poulin (1992b) compared the specificity of several parasite groups in Canadian freshwater fishes. This study confirmed that monogeneans are highly host-specific, even in freshwater, followed in degree of specificity by tapeworms and copepods (fig. 8.10).

As I just indicated, genomic research allows us to separate "cryptic" species that cannot be distinguished morphologically. At least this is possible when we are dealing with species whose geographic ranges overlap—that is, for whom there is no physical obstacle to sharing genes.

For example, Jousson, Bartoli, and Pawlowski (2000) had to resolve the following question. In the Mediterranean live three fish species in the genus *Diplodus*: *D. sargus*, *D. vulgaris*, and *D. annularis*. All carry trematodes in their guts. When Jousson, Bartoli, and Pawlowski began their research, it was believed that the three *Diplodus* species all harbored two trematode species identified as *Macvicaria crassigula* and *Monorchis parvus*. Each of these two trematode species was therefore assumed to be not very specific, given that each infected all three fish species. By analyzing certain parts of the genome, Jousson, Bartoli, and Pawlowski determined the phylogenetic relationships among a series of specimens of parasites gathered in the three host fish species (fig. 8.11). The analysis was conducted for both *Macvicaria* and *Monorchis*. The phylogenetic trees (obtained by a method called neighbor-joining) showed that *Macvicaria crassigula* is in fact a complex of two species, one of which parasitizes *D. sargus* and *D. vulgaris*, whereas the other parasitizes *D. annularis*. Most surprising is that *Monorchis parvus* is also a species complex, of which one species parasitizes *D. sargus* and *D. vulgaris* and the other parasitizes *D. annularis*, just as for *Macvicaria*! The molecular analysis thus revealed the trematodes to be much more narrowly host-specific than had been suggested by simple morphological analysis. Jousson, Bartoli, and Pawlowski (2000, 781) concluded their paper with a hypothesis on the possible causes of speciation in *Macvicaria* and *Monorchis*: "surprisingly, the pattern of cryptic diversity is identical for *M. crassigula* and *M. parvus*, although these species belong to distantly related families. . . . This would suggest that the speciation of *Diplodus* parasites was driven by factors associated with the hosts."

Richness

Ecologists use the term *species richness* to talk about the number of different species encountered in a particular habitat or site. An example is the great species richness of tropical rainforests and coral reef ecosystems. Or we say that one island has greater species richness than another island, and so forth. The term is also used in parasite ecology.

Parasite richness can be analyzed at several levels, either taxonomic (some particular group of parasites) or geographic (some region). Parasite richness can be stated only approximately if the sample is not large enough; it is only beyond some threshold of sampling intensity, which varies with host species, that the known parasite richness can be considered representative of the true richness. It is of great interest to ask what causes the distribution of parasites among hosts such that some hosts have many parasites and others few.

At the ecosystem level, there are many hypotheses about traits of host species that render them more or less suitable for parasite species. Figure 8.12 (*right*) shows a number of hypotheses formulated about traits that might positively influence parasite species richness in a host species. These hypotheses are based on the following logical statements:

- Diet: the more diverse a host's diet, the more frequent the occasions for eating infective stages harbored by different intermediate hosts (vectors).
- Contact with water: the more intimate and prolonged the contact with water, the more a host species can contact parasites, because many parasites have aquatic infective stages.
- Position in the food chain: the higher a host species is in the food chain, the greater the concentration of parasites with complex life cycles.
- Accessibility: the more the host is exposed, the more vulnerable it is to attack; by contrast, if it has a refuge, it can be partly protected (this argument applies mostly to parasitoids and vectored diseases).
- Latitude: the closer the ecosystem is to the Equator, the richer it is in free-living species; it is reasonable to believe it is also richer in parasite species.
- Complexity: the more complex the host species, the more habitats it will have that parasites can use; for example, a mammal ought to have more parasite species than a copepod.

It is evident that some of these hypotheses contradict one another. For example, host species A may offer more microhabitats to parasites than host species B (possibly because the alimentary tract of A, or some other organ, is more complex), but it is also possible that species B has more contact with water than species A does.

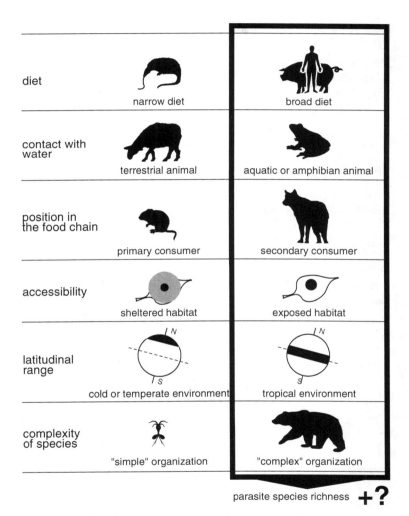

Fig. 8.12. Hypothetical determinants of parasite richness, envisaged for any collection of hosts.

Figure 8.13 shows the same types of prediction, but here in more precise versions because they are limited to a single host taxon; fishes are used as an example. The tailored hypotheses are as follows:

- Area: the greater the host geographic range, the more opportunities for it to acquire parasites from other host species.
- Gregariousness: gregarious behavior should facilitate transmission of parasites with direct life cycles and therefore limit the possibility that the parasites will be lost from a host species.
- Depth: fishes living near the bottom share this habitat with benthic invertebrates (molluscs, crustaceans, etc.) that are intermediate hosts for many parasites.
- Migration: a fish species that frequents many habitats during its migration can contact different parasites in each of them, so total parasite richness will be increased.
- Species richness of the phylum: transfers of parasites between host species will be easier if the host species are related, so a species-rich phylum multiplies the possibilities for transfer.
- Size: the larger host body size, the greater the available space; parasite species may compete for species.

None of these predictions is always realized, and the result often differs between host taxa. It is not surprising, for example, that species richness of nematodes should be influenced by different factors than species richness of tapeworms. However, quite robust general tendencies can be seen. Among the host traits I have cited, body size, geographic range, and species richness of the host phylum are traits generally associated with high parasite species richness.

This last trait—species richness of host phylum—is perhaps most important. For instance, in a comparative analysis centered on macroparasites of Canadian freshwater fishes, Robert Poulin (1992b) showed that, when a parasite infects a host species belonging to a family with many species, specificity is generally less strict. Host spectrum increases as a function of the number of related host species.

But the parasitologist, with the present state of knowledge, remains modest. It is risky to predict parasite richness of a host species in a particular environment if no direct study has been conducted.

Ecosystem Space

Parasites "circulate" in the ecosystem. Some follow simple routes and others complex ones through life cycles in which unrelated hosts follow one another. One interesting and still poorly studied aspect of parasitology

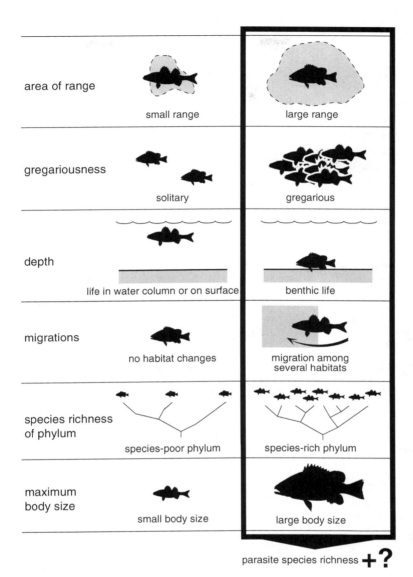

area of range	small range	large range
gregariousness	solitary	gregarious
depth	life in water column or on surface	benthic life
migrations	no habitat changes	migration among several habitats
species richness of phylum	species-poor phylum	species-rich phylum
maximum body size	small body size	large body size

parasite species richness **+?**

Fig. 8.13. Hypothetical determinants of parasite richness, envisaged for a related group of hosts—fishes, in this instance.

consists of reconstructing such routes in order to acquire key knowledge about the functioning of the ecosystem itself. The strategy is, if you tell me who parasitizes you, I will tell you whom you eat.

P. Bartoli's results in the coastal ecosystems of the western Mediterranean, thanks to multiple methods ranging from field observation through DNA sequencing, have allowed him to show how parasites are good indicators of ecosystem function and how they affect this function. Examination of 67 Caspian gulls (*Larus cachinnans*) in the Scandola Nature Reserve of Corsica yielded 15 different trematode species. Life cycles of 12 of them have been elucidated in the laboratory. Without exception, the birds acquire these parasites by ingesting metacercariae in their food. An inventory of trematodes provides much information on the birds themselves (Bartoli 1989).

First Datum: The Diet

Because of parasite specificity, the metacercariae of each trematode species are found in a clearly recognized intermediate host species. The very fact of finding the adult parasite in the digestive tract (or its appendices) in a gull shows unambiguously that the gull has eaten a certain type of prey. Figure 8.14 indicates schematically how the range of typical prey of the Caspian gull was deduced from the presence of different parasite species.

We notice that marine fishes constitute the major part of the *Larus* diet, as expected, but also that other, more surprising prey can be eaten, the strangest being snails and terrestrial insects. In fact, the analysis can be taken further because of the highly variable specificity of metacercariae with respect to their vectors.

Although the trematode *Galactosomum timondavidi* can be ingested with many fish species, by contrast, *Nephromonorca lari* is found only in silversides and *Condylocotyla pilodora* is found only in eels. We therefore have precise indications of the diet of these gulls. Better yet, it is reasonable to hypothesize that, to some degree, the frequency of a parasite will be correlated with the dietary habits of the gull. Consequently, the high frequency of *Nephromonorca lari* shows not only that silversides are eaten but also that they are probably particularly sought as prey.

Second Datum: Particular Needs

The abundances of parasites yield information on the degree to which the opening of the filters of encounter and compatibility differs between

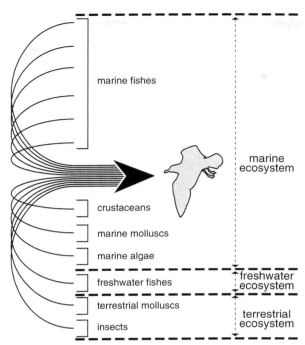

Fig. 8.14. Regular prey of the Caspian gull deduced from its parasites.

males and females. For the gull, Bartoli showed that the abundance of the trematode *Aporchis massiliensis* is much higher in females, and particularly in females brooding eggs. The explanation is that, during reproduction, the females eat huge quantities of kelp fronds, and it is precisely on these fronds that the metacercariae of this trematode are found. This dietary change is associated with the need for vitamins by the nestlings (Bartoli, Bourgeay-Causse, and Combes 1997). Moreover, all evidence suggests that the location of the parasite at the tips of the kelp fronds is a good example of localization in space. Therefore, although we cannot completely eliminate the possibility that the filter of compatibility is at play here (the females may be more susceptible to *A. massiliensis*), it is likely that the feeding behavior during the brooding period opens the filter of encounter to this parasite (fig. 8.15).

Third Datum: Migrations

Another type of information arises from one trematode, *Gymnophallus deliciosus*, that has been found (6 adults) in the bile duct of one gull on

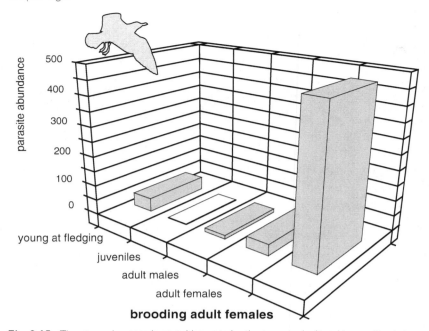

Fig. 8.15. The strong increase in parasitism rate by the trematode *Aporchis massiliensis* in brooding female Caspian gulls. Abundance is prevalence (percentage of individuals parasitized) multiplied by mean intensity (mean number of parasites per host individual), or the number of parasite individuals per 100 host individuals (see fig. 8.1). The fact that young are already parasitized at fledging is explained by the contaminated food regurgitated for them by the females.

Corsica. This is a parasite whose life cycle unfolds only in northern latitudes (North Sea, Irish Sea, Barents Sea, Baltic Sea). This fact indicates that the gull captured on Corsica had, during the previous weeks, visited cold regions. At a more local scale, the roster of trematodes gathered at Scandola (on the west coast of Corsica), along with exact knowledge of the distribution of intermediate hosts, showed that the same gull can harbor parasites that could have been transmitted at scattered points around the coast of the island.

Planetary Space

All species are sensitive to the environment, and because the earth is a veritable mosaic of different habitats, different species have very different distributions. Further, it is these distributions that, to some degree, limit competition between species.

For parasites, as I have shown earlier, the first level at which spatial heterogeneity can be detected is that of host species and even organs within the host. This level is often called the *microhabitat level*. However, the habitat also plays a role at a much larger scale, that of the *macroenvironment*. The macroenvironment can play a direct role (for example, the temperature may not suit the free-living stage of some species) or an indirect one (for example, the temperature may not suit some host species necessary for completion of the life cycle). Of course, it is not always easy to separate direct and indirect causes.

The Periwinkles of Arctic Shores

Periwinkles are small marine prosobranch gastropods that are cosmopolitan but particularly abundant on shores of cold seas. There they occupy the intertidal zone and are intermediate hosts for many trematode species, the majority of which mature in seabirds (for example, gulls and eiders). *Littorina obtusata* and *L. saxatilis* have largely overlapping distributions. However, *L. saxatilis* tends to locate a bit higher in the intertidal zone than *L. obtusata,* and it prefers rocky substrates, whereas *L. obtusata* is found on macroalgae. These are rather sedentary animals; their lifetime movements are within a radius of 2 to 3 meters, although they can live for ten years.

Russian researchers conducted extremely detailed studies on the parasite life cycles that incorporate these two species as hosts. Let me summarize results of Granovitch, Sergievsky, and Sokolova (2000) bearing on the geographic distribution of these parasites in the White Sea, east of Finland and Karelia, partly above the Arctic Circle. The White Sea penetrates deep into Russia, forming several narrow bays that are icebound nearly half the year. At the base of one of these bays (Kandalaksha Bay) are the study sites of these scientists, on the Northern Archipelago. There, over a dozen kilometers, are scattered rocky islets, the largest of which are 3 kilometers long, while the smallest extend only several hundred meters. This is a nature reserve where birds are particularly abundant. The sites where periwinkles were sampled are located on the banks of these islands.

Figure 8.16 depicts some of the results. The three graphs correspond to three species of *Microphallus*: *M. piriformes, M. pygmaeus,* and *M. pseudopygmaeus.* In each graph, the sites are arranged in decreasing order of rates of parasitism (prevalence) of *L. obtusata* (from left to right), yielding the front curve. The rear curve is the rate of parasitism of the second periwinkle species, *L. saxatilis.* We notice two remarkable facts: on the one

Fig. 8.16. Parasitism of two species of *Littorina* (gastropod molluscs) in locations on the shores of the White Sea. In each diagram, locations have been arranged by decreasing parasitism rate in *L. obtusata*, in order to determine visually if the same tendencies obtain in *L. saxatilis*. (Data from Granovitch, Sergievsky, and Sokolova 2000.)

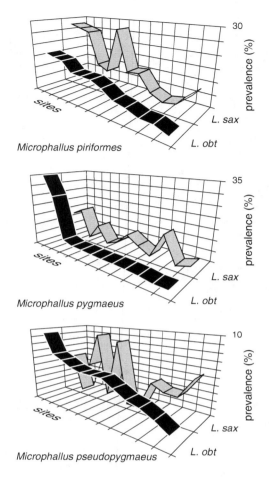

hand, parasitism varies greatly from one site to another (although the sites are often within a kilometer of each other). On the other hand, there is no correlation between the parasitism rates on the two *Littorina* species (the rear curve is not congruent with the front curve; this is especially evident for *M. pygmaeus*, for which the two sites with strong parasitism of *L. obtusata* show nothing of the sort for *L. saxatilis*).

Because the study covered 10 years and each site was not sampled every year, the observed differences can be ascribed to either spatial or temporal variation (a yearly follow-up study in certain sites showed that temporal variation was not negligible). However, on the basis of their observations, the scientists believe the sites with high parasitism rates mainly coincide with sites of seabird reproduction. As for the observed differences between the two periwinkle species, they are probably explained by the notably different ecology of the two species and possibly by different compatibilities for the different *Microphallus* species.

This inquiry demonstrates that, in sedentary hosts such as *Littorina*, parasitism rate can vary greatly from site to site, depending on either the physical environment (because the two periwinkle species do not occupy exactly the same coastal zone) or the biotic environment (because seabird reproduction is concentrated on certain sites along the shore).

The White Sea study, however, was conducted at a very fine scale, on the order of several kilometers, and we understand that multiple factors can influence local variation in rate of parasitism of periwinkles. Factors such as tidal currents, wave exposure, and slope may play a role. It is therefore interesting to analyze parasitism at a large scale, one able to reveal more general causes by neutralizing the background "noise" generated by multiple factors that are hard to identify. A study by Galaktionov and Bustnes (1999) fits this prescription. This study still focuses on parasitism of these periwinkles but this time uses sampling stations scattered on the coasts of the Barents Sea, in the north of Russia, Finland, and Scandinavia. Galaktionov and Bustnes examined 26,000 molluscs.

One of their most interesting results concerns the relationship of the frequency of trematodes present in periwinkles to the presence or absence of free-living stages (miracidia and/or cercariae). In a life cycle without a free-living stage, the mollusc ingests the egg (there is no swimming miracidium), and the cercariae remain in the mollusc in which they were produced (there is no swimming cercaria). In a cycle with a free-living stage, at least one of these two larval forms is free-living and thus confronts the external environment. All 13 trematode species studied in this research are species whose adults are found in seabirds. The sampled zones run from Novaya Zemlya (Russia) in the east to the region of Tromsø (Norway) in the west. Each zone includes many sampling localities, a procedure that tends to damp the influence of local factors. The White Sea, where the previously described research was conducted, is located approximately in the middle of the area encompassed in the present study.

Figure 8.17 shows that, in *L. saxatilis* as in *L. obtusata*, only trematodes without free-living stages are present in Novaya Zemlya, whereas trematodes with free-living stages are progressively added toward the west. The total number of trematode species increases significantly from east to west, from 5 at Novaya Zemlya to 12 at Tromsø. Climatic conditions are

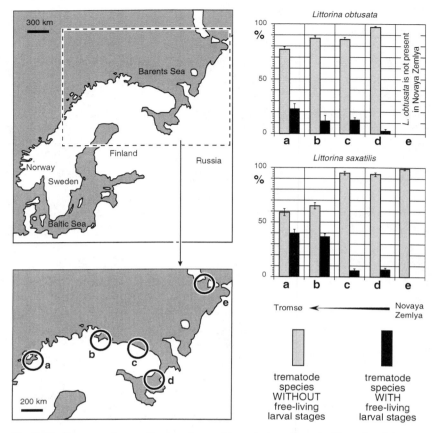

Fig. 8.17. Frequency, in parasitized molluscs, of trematodes with and without free-living stages in five sampling stations from Novaya Zemlya to northern Norway. (After Galaktionov and Bustnes 1999, modified.)

rigorous throughout the study area. However, the climate in the east is polar (in Novaya Zemlya, water is frozen from October to June) and in the west is relatively temperate (the influence of Atlantic currents reaches Tromsø). The correlation between climatic rigor and the absence or rarity of life cycles with free-living stages led the scientists to conclude that transmission of free-living, swimming stages is impossible in the coldest water, frozen most of the year.

This inquiry shows that parasitism of a single host species can vary greatly from locality to locality, influenced by identifiable factors.

The Alga That Kills Parasites

Parasites (especially those with complex life cycles) that require several host species can be excellent environmental indicators. To date, this approach has been the object of only a few studies. The reason is that such research is difficult and few specialists are willing to risk it.

Several threads, however, converge to suggest that parasitism is sensitive to factors such as seawater salinity, waste pollution, heavy metal concentration, and thermal pollution by nuclear power facilities. For example, various studies reveal that pollution entrains a weakening of parasitism (possibly because free-living stages or intermediate hosts are disfavored) or, on the contrary, increased parasitism (possibly because host immunity is compromised). It has even been shown that certain intestinal parasites are able to accumulate heavy metals in concentrations so substantial that they can, paradoxically, protect their hosts; in the rat, for example, the acanthocephalan *Moniliformis moniliformis* sequesters cadmium to concentrations up to 100 times that of its host tissues (see Sures, Siddall, and Taraschewski 1999 for a review of heavy metal concentration by parasites and Sures 2004 for a review of the use of parasites to monitor pollution).

Any variation in ecosystem composition, even over a short distance, can therefore produce strong variation in parasitism of a given host species. P. Bartoli and C.-F. Boudouresque (1997) have demonstrated this fact in a comparative study of parasitism of a wrasse, *Symphodus ocellatus,* in sites invaded by the famous "killer alga," *Caulerpa taxifolia,* and in uninvaded control sites. This alga, which has spread over the Mediterranean shelves for about twenty years and has been the subject of heated controversy (cf. Meinesz 1999), locally forms very dense meadows where it appears to decrease biodiversity, all the while leaving some sites uninvaded. The sampling for the parasitological inquest was conducted on the shores of the Alpes-Maritimes department, in a transect about 20 kilometers long.

A total of 134 *S. ocellatus* were studied at uninvaded sites and 131 at sites dominated by *C. taxifolia.* Six intestinal trematode species were found in these fishes. Because sizes of the two samples are very similar, in figure 8.18 I have summed the number of parasites collected, without analyzing the data. Although this type of inquiry must be confirmed (simply because it is always difficult to show that a correlation reflects a cause-and-effect relationship), it appears that parasitism is practically absent in the *Caulerpa*-infested sites, while, for at least three parasites, there are substantial parasitism rates in the sites without the alga.

At least two hypotheses can explain this result. Either the alga *Caulerpa taxifolia* is indeed a killer of intestinal parasites of *S. ocellatus,* or the *Caulerpa*

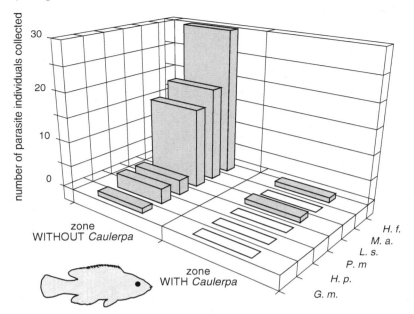

Fig. 8.18. The "killer" alga *Caulerpa taxifolia* causes loss of parasites of *Symphodus ocellatus* (data from Bartoli and Boudouresque 1997). *G. m., Genitocotyla mediterranea; H. f., Helicometra fasciata; H. p., Holorchis pycnoporus; L. s., Lecithaster stellatus; M. a., Macvicaria alacris; P. m., Proctoeces maculatus.*

meadows affect the intermediate hosts of these trematodes such that the parasite life cycles cannot be completed. The first hypothesis cannot be refuted because some plants, especially in the tropics, contain toxic compounds that repel predators and could also adversely affect parasites. The existence and distribution of these compounds support the "nasty host" hypothesis of Gauld, Gaston, and Janzen (1992), according to which the toxins would explain, for instance, why the diversity of tropical insect parasitoids is less than might have been expected. However, the second hypothesis, which posits an indirect effect on parasite life cycles, is much more likely. At least *Symphodus* should rejoice in the invasion by *C. taxifolia*!

Planetary Time

Exploration of the Planet

Parasites rarely leave fossils, so it is not easy to reconstruct their past movements on the planet—for example, to determine where a group arose, how its geographic range varied in the past, and to what extent it followed the migrations of its hosts. However, DNA sequencing has allowed the formulation of certain hypotheses on parasite movements on a global scale, starting with data from existing species. From sequences obtained for the same locus in different species, it is possible to construct phylogenetic trees using appropriate computer programs. The method is simple and based on the assumption that mutations occur at a steady rate and therefore the time since two species have diverged is proportional to how different their sequences are.

Parasites of humans are the object of the most thorough research, because this matter is part of the larger history of human diseases. One particularly well-studied group is the schistosomes, which I have already discussed in several chapters. Many schistosome species parasitize mammals and birds that have frequent enough contact with aquatic habitats that transmission can occur. Currently humans are affected in Asia by a species that is not very host-specific, *Schistosoma japonicum*, while in Africa humans host two more specialized species, *S. mansoni* and *S. haematobium*.[6] In South America, only *S. mansoni*, one of the two African species, is present.

What is the history of this group? First, Després et al. (1992) and Després, Imbert-Establet, and Monnerot (1993) showed that the African parasites of humans are closely related to parasites of either rodents (*S. mansoni*) or ungulates (*S. haematobium*) and that *S. mansoni* in South America is none other than the *S. mansoni* of Africa carried to the New World with the slave trade.

Later, Snyder and Loker (2000) and Lockyer et al. (2003) succeeded in analyzing DNA from a much greater number of schistosome species, parasites of mammals and birds from diverse regions. This analysis showed that *S. japonicum* and closely related species have as a sister group (on the phylogenetic tree) a cluster of species formed by an Asian parasite of cattle (*Orientobilharzia turkestanicum*) and by the African parasites *S. mansoni* and *S. haematobium*.[7] The analysis suggests that the genus *Schistosoma* was originally Asian and only secondarily invaded Africa, where it differentiated into several species, parasites of ungulates, rodents, and finally humans. In historical times, very recently, one species (*S. mansoni*) passed with humans to South America.

Exploration of the Ecosystem

I said that there was a beginning to every parasitism, that is, a passage to parasitism by a species that was originally free-living. I also said that parasites are most often quite specific to their hosts. Does this specificity mean that these associations are fixed indefinitely? Yes and no. Yes, if the time scale is short, perhaps on the order of a million years. This means that, in principle, parasites remain faithful to their hosts but that this fidelity has limits. In each generation, the infective stages of a given parasite can be led, for whatever reason, to "explore" the ecosystem—that is, to encounter different species of hosts than the one they normally inhabit.

If we can reconstruct, first the phylogeny of a group of hosts, then separately the phylogeny of a group of parasites associated with these hosts, it is possible to compare these two phylogenetic trees. We then notice that certain parasites evolve in parallel with their hosts, going so far as to "split" into two or more species each time the hosts themselves undergo a speciation; we speak of such a process as *cospeciation*. Johnson et al. (2003) have pointed out that, in cases in which hosts speciate but parasites do not, a likely explanation is that gene flow among parasite populations greatly exceeds that among host populations. Such gene flow would prevent the parasite populations from becoming sufficiently distinct that they would constitute new species. We also see that some parasites are quicker to try passages from one host lineage to another. (We speak of lateral transfers or "captures"; I mentioned such transfers to humans in the controversy closing chap. 4.)

On a scale of several million years, it would rarely be the case that no such captures occurred, whatever the groups of parasites or hosts of concern. Study of the frequency of such transfers has been greatly spurred beginning in the late 1980s by analysis of DNA of current species. If the branch tips of a phylogenetic tree of a parasite taxon are replaced by the hosts of the respective parasites, we can easily compare the topology of this tree with that of the phylogenetic tree of hosts.

When we compare phylogenetic trees of a group of hosts and their parasites, we observe one of three possible situations:

1. The trees are completely congruent, because the speciations have always occurred at the same time in the parasites as in the hosts. Evolution, including speciation events, has been parallel in the two taxa (fig. 8.19, *top*). This situation is rare in nature.
2. There is an overall resemblance between the phylogenetic trees, but certain parts are not perfectly congruent. This is the most frequent situation. The incongruences are the result either of parasite transfers between hosts that are themselves closely related or of errors in the analysis. We must always remember that the raw data for the

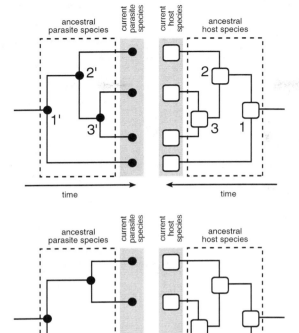

Fig. 8.19. Lateral transfer of parasites revealed by phylogenetic trees based on DNA sequencing. (*Top*) Congruence between phylogenetic trees of parasites (*left*) and hosts (*right*) suggests parallel speciation events; (*bottom*) lack of congruence suggests that at least some lateral transfers have occurred.

analysis are nucleic acid sequences that are but a small fraction of the entire genome.

3. Some differences between the phylogenetic trees can only be explained by surprising lateral transfers in which some parasites "jump" from one host group to another that is not closely related (fig. 8.19, *bottom*).

Figure 8.20 depicts a transfer clearly indicated by a comparison of the phylogenies of the parasites (monogeneans of the genus *Polystoma*) and their hosts (frogs and toads). Frogs and toads comprise two clearly identifiable groups, the archaeobatrachia and the neobatrachia. Molecular phylogenies show that *Polystoma* evolved in the neobatrachia, but we see immediately that there is an exception: species 12 (*P. pelobatis*) parasitizes an

222

Fig. 8.20. Phylogenies of polystomatids and their amphibian hosts, derived from combined analyses of several DNA sequences; branch lengths not considered.

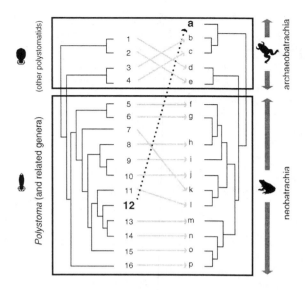

archaeobatrachian species, *a* (*Pelobates cultripes*). Now, the molecular phylogeny of *Polystoma* shows that *P. pelobatis* is very close to *Polystoma* found on neobatrachia and that its ancestors did not parasitize archaeobatrachia. This is therefore a case of a recent transfer from a common ancestor that lived on a neobatrachian species (Bentz et al. 2001).

Sometimes it is possible to deduce the ecological or behavioral factors that facilitated a transfer. For example, Page et al. (1998) showed that the phylogeny of lice of the genus *Dennyus* (mentioned earlier in this chapter) is broadly congruent with the phylogeny of its hosts, swiftlets of the genus *Aerodramus* in Indonesia. However, in several instances they found the same louse species on unrelated hosts. Such a species is *Dennyus simberloffi*, which normally parasitizes the black-nest swiftlet, *Aerodramus maximus*, but has also been collected on *Aerodramus fuciphagus*, the edible-nest swift, which is not closely related to *A. maximus*. The authors explain that "these two birds nest in close proximity . . . and it appears that this proximity has provided opportunities for *D. simberloffi* individuals to accidentally infest hosts other than *A. maximus*. [This] is evidence that swiftlet lice can disperse to foreign hosts, suggesting that they may have successfully colonized hosts in the past" (Page et al. 1998, 288).

As a general rule, lateral transfers occur when two free- living species (I will call them A and B) that were formerly kept apart by their habitats or behaviors become more similar in this regard, by whatever means. If species A carries a parasite, this parasite can then pass to species B. The convergence of habitat or behavior corresponds to opening the encounter filter. Of course, the fact that two free-living species share a certain lifestyle by no means guarantees that they will exchange parasites. It would still be necessary for the compatibility filter to be at least partly open, which would depend on the resources available in the new host and on the immune systems of the two host species, as well as on the ability of the parasite to adapt to the available resources and the new immune system.

Exploring the Host

In the course of geological time, parasites therefore explore ecosystem resources. Do they also explore the bodies of their hosts? Yes! A parasite contacting a host resembles a spelunker in a dark cave full of galleries radiating in all directions. When it is stopped by an obstacle, it can sometimes even remove the obstacle (just as a spelunker does), using enzymes where the spelunker might use explosives.

If the spelunker already knows the cave, a further analogy is possible, for the parasite also possesses a "map" of the host. If it is a question of a parasite of humans, for example, the parasite "knows" that, even though it has never "seen" the individual it has entered, the individual has a circulatory system, an intestine, a liver, and so forth. (This predictable environment is called the "third environment" by Sukhdeo [2000], for whom the first and second environments, respectively, are the aquatic and terrestrial ones.) It is this knowledge inscribed in the genes that permits the parasite, whatever its mode of infection, to find the organ that suits it by detecting and following biochemical gradients.

Just as the cave explorer often gets lost, however, so does the parasite. In most cases this risky exploration is lethal. However, given enough time, it is possible that an individual parasite wandering outside the normally permitted path happens to carry mutations allowing it to establish in a different organ. With a bit of luck, such an event may give rise to a parasite lineage specialized for this new microhabitat.

A family of monogeneans, the Polystomatidae, escaped from their normal fish hosts and conquered terrestrial tetrapods (amphibians, turtles, and a hippopotamus species). This host diversification was accompanied by an exuberant diversification of habitats. As depicted in figure 8.21, the Polystomatidae, originally parasites in the skin and gills of fishes (the ancestral hosts), have successfully explored the urinary bladder, esophagus, and oral cavity of their tetrapod hosts. The figure also shows, as dotted lines, the remarkable migrations that some parasites undertake before

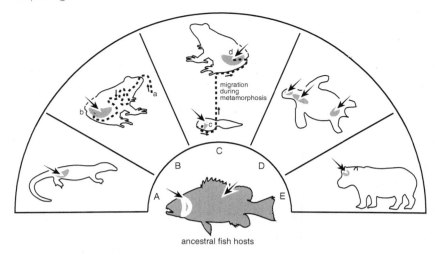

Fig. 8.21. Habitat change by monogeneans in the course of evolution. Although primitively ectoparasitic on fishes (skin and gills), some species have become mesoparasites in tetrapods. (A) In the giant salamander of Japan, *Pseudopolystoma dendriticum* inhabits the bladder, and the ciliated swimming larva (oncomiracidium) penetrates it directly. (B) In toads from North American deserts, *Pseudiplorchis americanus* also parasitizes the bladder, but the swimming larva enters the toad by its nostrils. (C) The most complicated transmission mode is that of *Polystoma integerrimum*, a parasite of the European common frog, *Rana temporaria*. The oncomiracidium attaches to tadpole gills and loses its ciliated cells; later, when the frog metamorphoses, the young monogenean emigrates to the bladder, where it matures in three years. There is synchronization of the host's egg laying and that of the parasite (both lay eggs once a year, in springtime). Moreover, if the oncomiracidium attaches to a very young tadpole, it develops in several days into an egg-laying form (morphologically different from that in the bladder!) that lays its eggs in situ but dies at metamorphosis. (D) In freshwater turtles, various species of polystomatids inhabit the eyelids, the esophagus, or the bladder and are believed to reach these sites directly. (E) In the hippopotamus, *Oculotrema hippopotami* is found under the eyelid and probably reaches this site directly.

arriving at their microhabitat. In *Pseudiplorchis americanus* (the line running from a to b), the swimming larva enters the adult amphibian host through its nostrils, remains in the lungs for several days, makes its way through the digestive tract, and finally reaches the bladder.[8] In *Polystoma integerrimum* (the line running from c to d), the swimming larva first attaches to the gills of a tadpole and migrates toward its bladder at metamorphosis. As we see, the conquest of terrestrial hosts entailed a passage from ectoparasitism to mesoparasitism.

Oddly, there are also monogenean parasites of rays and sharks (that is, aquatic hosts) that have become mesoparasites. Euzet and Combes (1998) proposed three hypotheses about selective pressures that might have entrained such a passage: the competition hypothesis (they avoid competition with other species located on the skin or in the gills); the predation hypothesis (they are escaping predation by cleaner fish that eat the ectoparasites of their clients); and the resource hypothesis (the host body fluids are more nutritious than the surface mucus).

In sum, exploration of new habitats is not rare in the history of parasites, or at least some groups of them.

Controversy: Inequality in the Face of Parasitism

Of what inequality am I speaking? Of the inequality that characterizes all host–parasite relationships at a spatial scale. This chapter has shown that, at one or more levels, parasites are never distributed homogeneously. Inequality is omnipresent in the world of parasitism.

The question What causes this inequality? is both exciting and difficult, whatever the scale under consideration. I will discuss two aspects of this problem and try to explain that parasites are very discriminating in exploiting the range of potential hosts in the biosphere and that host populations comprise individuals that differ greatly in their response to parasitism.

Why are parasite species generally quite narrowly specialized in their use of hosts? Why do parasites have so many analogs of free-living pandas and koalas?

Many hypotheses have been advanced to explain the degree of parasite specificity. I cite the two among them that seem especially well supported. The first hypothesis suggests that, for reasons of cost, parasites cannot be generalists. I term this the cost hypothesis. The second hypothesis suggests that parasites derive benefits from not being generalists. I term this the benefit hypothesis.

The Cost Hypothesis

Adaptations that allow a predator to eat prey of different species impose a cost, be it behavioral (for example, to have reactions that adapt a predator to capture prey that themselves have different behaviors), morphological (for example, to have teeth adapted to holding or chewing substances of different consistency), or physiological (for example, to have enzymes capable of digesting different sorts of tissues). This cost explains why there are limits to the "prey spectra" of all predators. Even for herbivores it is evident that no one species can survive equally well on all possible plants.

In parasites, the costs I just discussed exist, even if the forms they take are different. For example, the behavior of free-living stages cannot be adapted to any target species unless that behavior aids in detecting that target, attaching to the substrate, or penetrating the host integument. As for feeding in the environment offered by the host, this also requires biochemical machinery tied to metabolism that cannot accommodate substances that are chemically very different from one another. The problem of settlement in host organs also confronts many parasite species; I have already mentioned the lice of cave swiftlets that cannot invest evolutionarily in organs that can grasp feather barbules of all diameters. Similarly, suckers, claws, or hooks of monogeneans that parasitize fishes have sizes and shapes that correspond closely to the gill filaments to which they attach, whereas the scolices of tapeworms are highly adapted to the structure of the villosities of the particular part of the intestine in which they tend to settle.[9]

On this level, then, there are comparable limitations to specialization for free-living and parasitic species.

What characterizes parasitism is the addition of an extra constraint, and thus an extra cost. As I have noted several times, any host invaded by a parasite marshals against this non-self entity a battery of immune defenses; these are always present, although their effectiveness varies with the host species. Whatever strategy a parasite employs to survive in the hostile environment created by the host, this strategy generates a cost. For example, to manipulate the host phenotype to render it less aggressive requires the production of molecules adapted to induce this manipulation. Now, it is possible that no two species use exactly the same processes for their immune defense. It is clear, then, that the multiplication of species in a host spectrum entails a considerable increase in costs for a parasite. The sum of these costs can become greater than the benefit the parasite would reap from exploiting several host species.

"Objection," you might say. Heteroxenous parasites already know how to exploit host species that differ completely. Every trematode, for example, has at least the following genes: those that allow it to exploit a mollusc and to suppress its defenses and those that allow it to exploit a vertebrate and to suppress its defenses. This parasite, even if it is host-specific during a given stage of its life cycle, is in no way host-specific over the entire course of this cycle. This objection is completely justified (fig. 8.22).

It is even one of the paradoxes of specificity: many parasites are "horizontally" specific (the ability to move from one host species to another at a given stage in the life cycle) but not "vertically" specific (the ability to move from one host species to another throughout the life cycle). And it is certain that this vertical nonspecificity imposes a considerable cost because it obliges the parasite to construct twice as many different enzymes, twice as

two closely related molluscs exploited in the same life cycle stage

a vertebrate and a mollusc exploited at two different life cycle stages

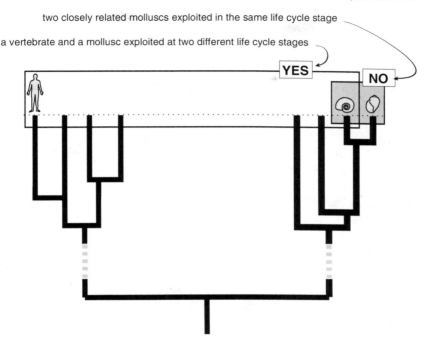

Fig. 8.22. The paradox of specificity.

many immunosuppressive molecules, and in a sense to construct two entirely different organisms (the trematode sporocyst or the redia inhabiting a mollusc bears no morphological, anatomical, or physiological resemblance to the trematode adult in a vertebrate). These costs must therefore by compensated for by benefits the parasites draw from heteroxenous life cycles. This deduction strongly validates the conclusion of chapter 3 to the effect that life-cycle complexity is advantageous and certainly not necessarily a handicap.

The Benefit Hypothesis

We know that every parasite population is fragmented into as many infrapopulations as there are host individuals. Therefore, at a given moment, genetic exchanges are possible only between individuals of the same infrapopulation. Moreover, parasites need genetic diversity as great as that of their hosts (see chap. 6) in order to produce a stream of new genotypes so they can remain competitive in the "arms race."

228   Let us now imagine a small parasite population. Assuming the size of this population remains unchanged, individuals of this population are more liable to be isolated from one another (that is, in separate infrapopulations) the greater the number of available host individuals. It is therefore in the interest of the parasites not to distribute themselves among too great a number of host individuals. Klaus Rohde (1994) first proposed this hypothesis to explain parasite specificity not for host species but for microhabitats within the host. In fact, we often observe very marked specialization for a particular organ or tissue, even when adjacent organs and tissues are not occupied by other parasites. To specialize in a particular host species and/or particular organ is to increase the probability of encountering a mate, especially if population size is small.

Combes and Théron (2000) suggested that there is a link between specialization, genetic diversity, and speciation (fig. 8.23). If we agree that parasites need genetic diversity to remain competitive in the arms race, we can argue, by enlarging Rohde's hypothesis, that specialization facilitates genetic exchanges, thereby increasing genetic diversity. But we can as well predict that this genetic diversity is advantageous in the conquest of new hosts and thus leads to a diminution of specificity. There is therefore an ongoing process, termed *alloxenic speciation* by Euzet and Combes (1980), that tends to fragment the host spectrum. Alloxenic speciation arises because, when a parasite exploits two or more host species, the differences in behavior or immune reactions of these species can lead to isolation of populations specialized for different hosts. For this isolation to occur, it suffices that there exists a polymorphism (that is, genetic diversity) for host preference. In other words, if certain individuals of one parasite population (denoted *a*) survive better on host species A, whereas others (denoted *b*) survive better on host species B, *a* parasites tend to encounter other *a* parasites, whereas *b* parasites tend to encounter other *b* parasites. If "gene flow" is therefore reduced and finally disappears between the *a* and *b* groups, a process leading to speciation is under way. Euzet and Combes have called this process *alloxenic speciation* to emphasize that the different (*allo*) hosts (*xeno*) are responsible. There is thus an ongoing restoration of host specificity and of the advantages it confers on parasites. It is very likely that the speciation by *Macvicaria* and *Monorchis* in fishes of the genus *Diplodus* reported by Jousson, Bartoli, and Pawlowski (2000) and described earlier in the chapter is of the alloxenic type.

Just as a change of hosts can isolate individuals of one parasite species and lead to speciation, so can change of microhabitat within one host species. Euzet and Combes (1980) have called this process *synxenic specia-*

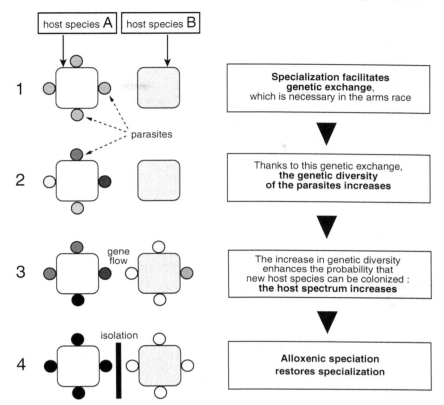

Fig. 8.23. The continual restoration of specificity (after Combes and Théron 2000, modified). Note that gene flow does not result from an exchange of adult parasites between two host species (once fixed in an individual host, adult parasites rarely leave it) but from the exchange of free-living stages in each generation in the external environment.

tion. It can explain why we sometimes find many parasite species that are congeneric (belonging to the same genus) in one host species but in distinct microhabitats.

Barton (2001, 325) writes that "speciation is the topic that links the interest of virtually all ecologists and evolutionary biologists," to which Schluter (2001, 378) adds, "speciation is one of the least understood features of evolution." The world of parasitism has escaped neither this intense interest in speciation nor the lack of knowledge about the precise mechanisms by which it occurs. Because a species is traditionally defined by its reproductive isolation from other species, the problem is to understand what processes can lead to this isolation. One such process, called *ecological speciation*, operates when natural selection in different environments (which, for example, might provide different resources) causes reproductive isolation. Ecological speciation is never easy to demonstrate (see Schluter 2001), but host–parasite systems, as these examples show, offer a very promising arena for its study. Owing to the frequent lack of mobility of parasites, host–parasite associations are especially valuable for analyzing sympatric speciation—that is, speciation that does not require geographic isolation in order for reproductive isolation to evolve (Hawthorne and Via 2001).

And Humans in All This?

We know (chap. 4) that the human species is unrivaled when it comes to lists of the parasite species it harbors.[10] We do not need detailed inquests to know that infectious and parasitic diseases are still, at the dawn of the third millennium, the main obstacle to the health and well-being of several billion people on this planet, mainly in poor tropical countries.

In what follows, I would like to show that amid the general prevalence of many parasites in human populations, individual humans are very unequal in the fight against parasitism. I have already discussed the general rule that parasites tend to aggregate in certain host individuals, and I have pointed out that this aggregation can be attributed to two heterogeneities—of genes and of conditions of life. This is equally true for the human species.

The Genes

As for differences in genetic constitution in the face of parasitism, the idea that certain individuals are more susceptible than others is old. However, only in the last ten years have strong demonstrations of these differences been available and the genes producing them been identified.

Alain Dessein and his colleagues have worked for many years in a small village, Caatinga do Moura, in the Bahia region of Brazil. This is a region where *Schistosoma mansoni* causes endemic schistosomiasis (Abel et al. 1991; Dessein et al. 1992). A river divides the village in two and is the center of many household and play activities. By analyzing number of parasite eggs per gram of feces, these researchers initially showed that most inhabitants were afflicted but also that there were great differences among individuals and families.

For example, the factors of age and frequency of immersion explain only 20–25 percent of the variance in infection level. The most heavily infected individuals were not randomly distributed within the village but rather clustered in certain families. Reinfection after antiparasite treatment was much quicker in some families than in others. These observations hint strongly at the existence of a resistance polymorphism in the population.

The authors then selected 20 families containing a total of 269 individuals and tried to relate level of parasitism to genetic factors. To this end, egg counts were not a good variable because, depending on age, sex, and personal habits, frequency of immersion in water varied.[11] Recognizing this bias, the scientists calculated an index, called E3, that is a transformed and standardized summary of the raw initial data. As shown in figure 8.24, the

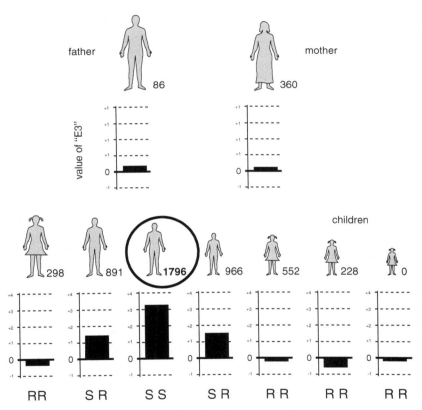

Fig. 8.24. Pedigree of a family from a Brazilian village showing that the corrected values (E3) of parasite load accord with the hypothesis of a major gene with two alleles (S and R) controlling individual resistance. SS, susceptible homozygotes; RR, resistant homozygotes; SR, heterozygotes. Numbers at the feet of individuals indicate the raw egg counts, in number of eggs per gram. Circle indicates the most heavily infected individual, assumed to be SS, and his egg load. Note that the family depicted here can give the impression that sex influences parasite load (males are more heavily infected than females); this pattern disappears when the entire group of families is analyzed.

results for one family from Caatinga do Moura indicate an inequality that leads to the reasonable hypothesis of a genetic origin, because other factors have been accounted for.

Comparing data from several families with mathematical models that predict the distribution of values of E3 under different genetic hypotheses, Dessein and his colleagues went further. They showed that their data accord best with a model of one major gene with two alleles, one (S) for susceptibility and the other (R) for resistance. This gene is called *SM1* (for *Schistosoma mansoni 1*). It is also possible to assign a genotype (SS, SR, or RR) to each individual. The figure shows the results in one family: two apparently heterozygous parents had one homozygous SS child, four homozygous RR children, and two heterozygous SR children. There is therefore probably a gene in Brazilian populations that largely determines success of infections, independent of length and frequency of bathing. This fact does not exclude the possibility that other genes play a role in resistance.

Subsequently, another gene, *SM2,* was identified that appears to control the severity of the illness in parasitized individuals. That is, certain but not all afflicted individuals suffer serious liver fibrosis even with relatively light infections (Dessein et al. 1999).

Conditions of Life

In the work just described, the researchers were at pains to neutralize the effect of the environment in order to focus on genetic effects. By contrast, an older study, by Hugues Picot and Jean Benoist (1975), aimed to reveal the importance of the environment and lifestyle, including cultural aspects.

This research was conducted in four villages on the island of La Réunion in the Indian Ocean: Bois-Blanc, Dos-d'Âne, Grègues, and Saint-Gilles. Among these villages, there are at least two heterogeneous factors. The physical environment (rainfall, elevation) differs among sites, and the composition of the human population also differs, with varying proportions of individuals of European, African, Madagascan, Indian, Chinese, and Comoros Islands backgrounds. These latter differences, in turn, generate behavioral differences.

Five main parasite species, with different transmission modes, were studied. Four were nematodes: *Trichocephalus* and *Ascaris* infect humans when their eggs are ingested, whereas *Ancylostoma* and *Strongyloides* infect by skin penetration in the humid, aquatic environment. The fifth species was a tapeworm with a heteroxenous life cycle, whose intermediate host is a beetle living in flour.

This study led to several conclusions. The most evident is that the four villages have very different infection rates for the five parasites (fig. 8.25). Picot and Benoist explain these differences by a mixture of climatic and behavioral differences. Bois-Blanc, without a pronounced dry season and with moist soils, is a propitious site for nematode transmission. Dos-d'Âne has a much lower prevalence of the two species that penetrate the skin, a consequence of a pronounced dry season. Grègues resembles Bois-Blanc climatically but the human behavior differs: Grègues has well-constructed houses, toilets, and medical services. Saint-Gilles is climatically similar to Dos-d'Âne, but its parasitism rate is reduced by better hygiene.

This first stage in interpreting the results thus emphasizes the two factors of climate and hygiene.

Picot and Benoist then studied the distribution of infections if the unit of analysis was households rather than individuals. They observed that

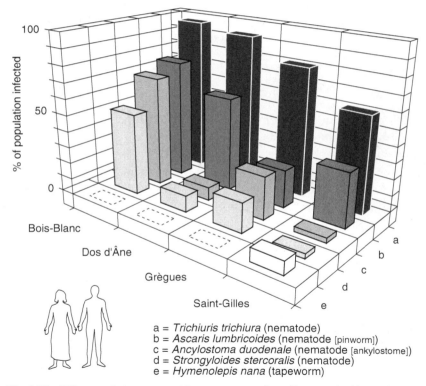

a = *Trichiuris trichiura* (nematode)
b = *Ascaris lumbricoides* (nematode [pinworm])
c = *Ancylostoma duodenale* (nematode [ankylostome])
d = *Strongyloides stercoralis* (nematode)
e = *Hymenolepis nana* (tapeworm)

Fig. 8.25. Differences in human parasitism rates among four villages on the island of La Réunion, mainly explained by cultural traditions.

234   infections were almost always aggregated in certain homes. For example, at Dos-d'Âne, 3 families of 71 contained 20 of the 38 infections of *Strongyloides*, whereas at Grègues, 5 of the 36 families constituted 23 of the 32 infections of *Ascaris*. This process of aggregation indicates that certain families live close to an infection source that has slight effect on other inhabitants of these villages. For the nematodes, this source can be a small body of water; for *Ascaris*, this could be in the immediate vicinity of the house.[12]

Finally, and probably most instructively, these researchers considered parasitism rates in light of traditional behaviors of different communities on the island. The most striking aspects of their report concern the Indian community, whose households manifested fewer cases of parasitism than did those of the other communities. "In their domestic lives, the Indians of Réunion have adhered to many rules and customary behaviors of their country of origin. These rules have even acquired a sacred value and are transmitted to children along with religious values. The separation of the pure and the impure, of the clean and the dirty are explicitly valued. There are ritual hand-washings, frequent baths, and shoes are considered 'dirty.' These practices establish a barrier between humans and the environment, even when the standard of living is low" (Picot and Benoist 1975, 245).[13]

Although genetic factors influencing resistance and susceptibility surely exist in the human populations of La Réunion and were not accounted for in this research, this study nonetheless illustrates the degree to which transmission of certain parasitic diseases depends on the environment. By emphasizing the importance of hygienic behavior, Picot and Benoist demonstrate that, in many countries, amelioration of the conditions of life by better education and a lessening of the precariousness of existence suffices by itself to reduce considerably, and possibly to eradicate, an important part of human parasitic diseases.

Emerging Diseases and the Future Arms Race

The human species is one fruit (among millions and millions) on the tree of evolution. This *Homo sapiens,* represented by a small community of clawless primates somewhere in the African savanna, without fangs or great strength, attacked the entire living world, destroying here, subjugating there, becoming the imperialist colonizer of the tree it is part of.

Paradoxically, although humans today have eliminated or relegated to nature preserves the predators that terrorized them throughout prehistory (and similarly treated other groups of hominids that bothered them), many infectious and parasitic diseases still threaten them. Having fallen into the magic potion prepared by the Red Queen, parasites have resisted humans and constitute an incongruous enemy bastion in the triumphalist tableau that the high-tech era proudly paints of itself.

In this battle against parasites, humans run up against the laws of natural selection. How did they discover them? By the Renaissance era and even before, some enlightened souls understood that life changes with time. However, it was really Jean-Baptiste Lamarck (1744–1829) who first elaborated the concept of species transformation. Lamarck had to deal with many problems, and in the middle of the nineteenth century there were still as many scientists who believed that species were unchanging as there were transformationists. If transformationism was difficult to establish as a dominant paradigm, it was mainly because its adherents could not propose

Parasites can hardly be receiving more attention than they deserve.

◊

C. A. TOFT (1991)

a credible mechanism to explain how living organisms become progressively modified. Lamarck had suggested that the changes acquired by individuals during their lives become hereditary, but he could not provide a coherent explanation of how this could occur.

It was then, in 1859, that Charles Darwin adopted, sometimes verbatim, Lamarck's transformationist ideas; however, he added an explanatory mechanism not based on the inheritance of acquired characteristics but on natural selection. For Darwin (as I showed in chap. 7), random, hereditary variations were continually produced in traits of all species, and among individuals, only the most "fit" (that is, the best adapted to the environment) survived, or at least reproduced more than others. I have already shown how many arguments Darwin could have marshaled from observations on interactions between parasites and their hosts if knowledge of parasites had been more advanced in his day.

A century and a half after Darwin, no one in scientific circles disputes that the history of life, spreading out over 3.5 billion years, consists of an infinity of transformations. However, the theory of evolution has not stopped at that point. Throughout the twentieth century, the processes that allowed the passage from the first living molecules through present-day vertebrates have been the object of hypotheses and counterhypotheses, all constantly enriched by new discoveries and interpretations.[1]

When I discussed Leigh Van Valen's Red Queen hypothesis (chap. 6), I showed that any increase in the fitness of any species modifies the environment of other species in that habitat and forces them to adapt. This adaptation, in turn, generates a change in the environment of the first species, which therefore undergoes another bout of selection ad infinitum; the only requirement is that the genetic diversity of each species be continually renewed by mutations (in the broad sense) in the genome. As Matt Ridley (1994) said, "any evolutionary progress will be relative."

The Red Queen hypothesis, however, can lead to two hypotheses that are both incorrect: on the one hand, that changes occur very regularly through time, and on the other hand, that different species are equally equipped to engage in the arms race.

◊

The question of the regularity of changes, or in other words, the gradualism of evolution, is one of the most controversial of the last half century.

The fundamental concept of the neo-Darwinian synthesis (that is, Darwinian theories revised and corrected in the light of genetic discoveries) is the gradualism of transformations. For the adherents of the theory that still prevailed in the 1950s, mutations induce only small changes, and it is the accumulation of these small changes that causes all the changes that have occurred in the history of life. Applied to minor traits like the size of a parasite or the spatial configuration of a membrane molecule, such a mechanism is easy to imagine and accept. However, when it is a question of explaining the origin of the major organizational strategies that characterize large, distinct taxa (such as molluscs, arthropods, annelids, echinoderms, or vertebrates), gradualism is no longer such a cogent explanation. Moreover, gradualism is not always apparent in the fossil remains of species.

Niles Eldredge and Stephen J. Gould, among others, argued vigorously beginning in 1972 against the proposition that all evolutionary change is gradual. According to them, phases of stagnation during which species change very little alternate with phases of rapid change, the latter occurring in small populations.[2] The fact that major changes occur in small populations suggests an important role for random "genetic drift." That is, in a population with few individuals, purely random changes in gene frequency can cause a gene (and thus a trait) to become more frequent than others, even if its fitness or selective value is no better.[3]

Physical events (continental fragmentation and drift, volcanic eruptions, glaciation and glacial retreat, etc.) are among the factors that, over relatively short periods, have produced major environmental changes to which certain species have responded evolutionarily with spectacular advances. The discovery of homeotic genes spurred a burst of research on the biology of these advances. Homeotic genes, coding for proteins that activate or suppress other genes, regulate the developmental stages of multicellular organisms. Much evidence suggests that homeotic genes originated toward the end of the Precambrian era, 700–800 million years ago—approximately when highly organized animals arose. It seems as if homeotic genes allowed and controlled the construction of increasingly complex organisms. A sudden increase in the number of these genes characterized the emergence of vertebrates about 400 million years ago.

Parasitologists are frustrated by not having participated in a major way in discussions about evolutionary tempo. Very rare fossil parasites have been identified, but they show only that parasitism is old. They do not allow us to date exciting developments such as, for example, the series of changes that led free-living tapeworm ancestors to evolve into present-day tapeworms, with their scolices armed with various attachment organs, their reproductive apparatuses repeated nearly to infinity, and their digestive tracts totally gone.[4]

The idea that acceleration phases separate periods of near stagnation does not contradict either the theory of evolution in general or the Red

238 Queen hypothesis. It only superimposes an uneven tempo on the underlying phenomenon. At the scale of an association between a parasite and its host, it is possible that the sequence of reciprocal "inventions" of the adversaries, tending to cancel one another out, usually occurs in a gradual way, although it may be punctuated by periods of more rapid evolution.

The evolution of hominids is one of the rapid events in which natural selection appears to have raced. In several hundred thousand years, it produced a brain of extreme complexity, associated with a dexterous hand and with language. This association produced an unprecedented revolution, the possibility of transmitting acquired characteristics. In humans, in effect, transmission of acquired knowledge from generation to generation became possible; a Lamarckian process took over from Darwinian mechanisms.[5] Lamarckian transmission produces change much more quickly than does natural selection. Nowadays, the greater and greater accumulation of knowledge follows an exponential curve.

This revolution directly affects relationships between humans and the parasite world, as the following section shows.

◊

The metaphor of the Red Queen running with Alice in a landscape that remains unchanged should not make us forget that the Queen and Alice are not running against each other; rather, they both run "against" the landscape, which represents "the others." When we apply the hypothesis to a host–parasite relationship, we can therefore have "Alice + the Queen" represent the host and posit that they run against a moving landscape that represents either a particular parasite or the community of all parasites of that host species.

Is the race a fair one? No, for several reasons, practically all of which favor the parasites. As I showed in chapter 1, parasites have (1) greater mutation rates than their hosts, (2) shorter generation lengths than their hosts, and (3) larger populations than their hosts. These three features guarantee greater genetic variability. Of course, hosts have responded to parasite weapons. Their genomes have quantitatively and qualitatively substantial resources that allow them to produce great genetic variation.

Nevertheless, if we return to the metaphor of Alice's running just to keep up with the landscape,[6] we have to imagine that Alice always lags a few steps behind it. This is the evolutionary lag that I discussed in detail in chapter 6. This lag exists in every species, whatever its environment. In fact, even when natural selection is fast, it can only fall behind the environmental change to which it is responding. The need for a minimal time necessary for natural selection to act means that if the environment changes, the adaptation of a species to that change is always late.

When we perform experimental infections with a particular parasite strain, local host strains, and host strains of the same species but from geographically distant regions, we say that there is local adaptation of the parasite if the parasite succeeds better in infecting the local host than the host from elsewhere. If, on the contrary, a particular host strain resists local parasite strains better than strains of the same parasite species from other regions, we say there is local adaptation of the host.

This type of experiment almost always shows that the parasite is better adapted locally than the host, which fits perfectly with the idea that parasites generally evolve more quickly than their hosts. For example, when molluscs (*Oncomelania hupensis*) from the Szechuan (Sichuan) region of China were exposed to infective stages (miracidia) of *Schistosoma* collected either locally or 1,200 kilometers away in the province of Anhui, the local parasites infected local molluscs in greater numbers and more persistently than did the distant parasites. The same result obtained (better infection of local molluscs by the local strain) when the experiment was run using molluscs from Anhui (Xia, Jourdane, and Combes 1998).

As for infectious and parasitic diseases of humans, by contrast, we have the impression that local adaptation favors the host: a human population can acquire a high degree of resistance to a local parasite, while a distant population, which has never "seen" the parasite, is highly vulnerable to it. (To see this point, it suffices to recall the history of the French expedition to Madagascar and the ravages of Native Americans wrought by "new" diseases brought by Europeans, which I have already discussed in chap. 6.)

In fact, we should not deduce from these examples that the human capacity to adapt to our parasites exceeds the capacity of the parasites to adapt to us. The relatively limited virulence of local parasites reflects an adaptation of the parasite rather than a human adaptation simply because parasite strains that are too virulent are selected against. As we know, a parasite that kills too many of its hosts or kills them too quickly lowers its own probabilities of survival and transmission. Parasites are able to adapt very quickly, and their degree of virulence evolves toward the value that assures the maximum reproductive success.

◊

For little more than a century now, a new form of struggle between parasites and humans is under way. For a clash initially between the genes of each, we have substituted a clash between the genes of one and the culture of the other. It is in this arena that the revolution fostered by Lamarckian transmission of knowledge truly flowers: through culture, humans can hope to redress the evolutionary lag and—why not?—end by being quicker than the parasites. The cultural transmission of acquired knowledge relegates the slow advance of natural selection to the shadows. Disciplines such as molecular biology, genomics, and information science have in just a few decades accumulated an impressive roster of discoveries and developments. The extremely weak information emitted by parasite infective stages (see chap. 4), considerably amplified by technology (optical and electron microscopy, chemical and genomic analyses), is exploited to close the filter of encounter as well as the filter of compatibility. The cultural investment entails both behavior (procuring clean water, cooking food, avoiding contagious disease, disinfecting objects, etc.) and direct battle against established pathogens by artificially modifying the battlefield, either a priori (vaccination) or a posteriori (therapy).

Infectious and parasitic diseases of humans and domesticated animals, even of cultivated plants, are thus the theater of a new coevolutionary process, because the parasites have been confronted, for a little over a century, not only with weapons produced by genomes but also weapons produced by culture. In some instances, cultural weapons have proven decisive: smallpox has been eradicated from the planet (except for vials maintained in Atlanta and Moscow), polio and diphtheria have disappeared from western Europe and the United States, pork tapeworm is only a bad memory, and many parasitic diseases have been mastered in genetically modified organisms. But these victories, important as they are, should not hide the fact that, in many instances, answering thrusts have already been selected for in parasites. Under the pressure of antibiotics, resistant bacterial strains arise constantly, and even the malaria agent, *Plasmodium falciparum*, required only a few decades to find a genetic response to the famous Nivaquine (chloroquine sulfate) and molecules derived from it.[7] The rapidity of natural selection in parasite genomes means that parasites are still redoubtable adversaries. Only if continually advancing knowledge produces "human" weapons more quickly than genetic mutations produce parasitic weapons can our victory remain intact.

No one doubts that the rapid lengthening of the average human lifespan through the nineteenth and twentieth centuries is closely correlated with biological discoveries. This fact implies that henceforth humans should rely more on their intelligence than on their genes to fight infectious and parasitic diseases.

With the appearance of culture in *Homo sapiens* and the accumulation of acquired cultural traits that culture allows, the struggle against all types of parasites became one of the products of humans as engineers of themselves. The edge in the race to adapt has perhaps, in just a few decades, shifted to humans.

Emerging Human Diseases

What Is an Emerging Disease?

Just as medical advances have caused the decline of many diseases, common sense would seem to dictate that this process will continue, so the frequency of infectious and parasitic diseases should logically tend toward an asymptote of zero. It has therefore been shocking to observe that, not only have "classical" diseases such as malaria not declined, but also a number of newly discovered pathogens are causing new—termed "emerging"—human diseases (Lashley and Durham 2002).

The expression *emerging diseases* actually designates three distinct groups of infectious or parasitic diseases.

First are diseases that are only "reemerging," that is, that have long been known but, having declined greatly in frequency, have been more or less forgotten. This is how we interpret several deadly epidemics of the last 20 years: purpuric fever in Brazil, caused by *Haemophilus influenzae;* hemorrhagic colitis from North America, caused by *Escherichia coli;* and several types of meningitis. In many instances, a sudden change in the severity of an infectious or parasitic disease is caused by genetic changes in known pathogens. Each time a mutation allows a pathogen to be transmitted more easily or to exploit its host better, there is an increased probability that natural selection will favor this mutation. Some pathogens routinely undergo mutations that occasionally greatly increase their impact and cause an outbreak that therefore seems to be an emerging disease. The best-known example is the influenza virus, which can be relatively benign or, on the contrary, highly malignant, as was the case at the beginning of the twentieth century, when the "Spanish influenza" epidemic of 1918–20 caused tens of millions of deaths.

In some instances, the genetic change responds to some therapeutic intervention. Thus, I have already noted how, in many pathogens, the intense selective pressure caused by the use of antibiotics has quickly generated the appearance of new, resistant strains. Among bacteria, such mutations most often affect the surface proteins that play a role in binding with antibiotics. We speak, for example, of a "decreased affinity of penicillin-binding proteins" among the *Streptococcus* bacteria that cause pneumonia or the *Neisseria* bacteria that cause meningitis (see Temime et al. 2003). It is not unusual for factors particular to the pathogen and to the behavior of the host to combine to enhance the impact of a disease. For instance, tuberculosis, which we had believed to have nearly disappeared, has for the last 15 years caused up to 3 million deaths annually. It is even possible that tuberculosis is currently the leading cause of death owing to infectious or parasitic disease in the world. The pathogenic agent *Mycobacterium tuberculosis* has undergone genetic changes that have allowed selection for resistance to many antibiotics. But human behavior has also played a role, especially through movement. In Iceland, for example, the number of strains of *M. tuberculosis* resistant to antibiotics has risen in a few years from 0 percent to 20 percent (in 1992). It is believed that such strains have been more common in countries farther south, where many Icelandic families spend their vacations. At the same time, the spread of these strains has been aided by the fact that 80 percent of the children in the capital city, Reykjavik, attend nursery schools or day-care centers. These changes in human behavior could by themselves, without genetic changes in the pathogen, have upset the previous equilibrium. Everyone remembers the toxic shock syndrome associated with the use of certain tampons by women, which was caused by a particular clone of *Staphylococcus aureus,* which currently exists among healthy women but does not proliferate in a pathogenic manner.

A second type of emerging disease is the "locally" emerging disease—that is, a disease formerly known in a different region of the planet. Earlier I wrote about the catastrophic impact on the health of native peoples that movements of a few dozen or a few hundred people can have when they invade new regions; consider the devastating diseases that ravaged Native Americans and Polynesians. In fact, we have good knowledge only of recent events of this sort. It is very likely that such movement of pathogens has occurred since prehistoric times. Many specialists in human prehistory now think that the evolution of hominids resembles a bush more than a single branch, so that several species of *Homo* coexisted simultaneously. By natural selection, some of these species were able to evolve resistance to a particular pathogen. One can readily imagine that, by carrying these pathogens to previously unexposed populations, these resistant species could have wrought catastrophes similar to those visited by Europeans on Native Americans. We can hypothesize, for example, that the rapid disappearance of Neanderthal man on contact with modern humans might have been caused by bacteria and/or viruses. Today, molecular biological techniques applied to ancient human remains, such as mummies, have detected the human transport of various pathogens in the course of migrations (Li et al. 1999). Nowadays, other movements, those of domesticated or wild animals, present similar risks of locally emerging diseases for native faunas. A consequence of movement of animals such as bats or the small Indian mongoose is that viruses, such as the one that causes rabies, have generated emerging diseases in the recipient ecosystems.

The third and last type consists of the "true" emerging diseases, diseases that are really new for a particular host species. In other words, these are what, at several points in this book, I have called transfers. These transfers are part of the normal evolution of parasitism, and we should not be surprised that, in the course of human prehistory and history, such diseases have emerged in all epochs (McKeown 1988). Emerging diseases are in no way a new phenomenon. As I have shown, schistosomiases have been emerging diseases in the past, and it is more than likely that changes in human behavior were responsible for transferring schistosomes (see Cockburn 1971; Combes 1991b). The arboreal lifestyle of our distant ancestors was hardly favorable to this type of disease, which is transmitted through contact with water. Studies of mummies show that schistosomiasis was already a plague in ancient Egypt, where agriculture and irrigation were highly developed (Contis and David 1996). Similarly, the tuberculosis pathogen that I previously discussed, *Mycobacterium tuberculosis*, probably originated from *Mycobacterium bovis*, found in livestock, and was passed to humans about 10,000 years ago as a consequence of rearing cattle. Domestication is always a risky practice; for instance, it has recently been shown that *Baylisascaris procyonis*, a roundworm that infects raccoons, is found increasingly frequently among humans in the United States (Sorvillo et al. 2002). Raccoons, which become infested by ingesting the eggs of the parasite, have increasingly become "peridomestic animals" living in close proximity to humans. When a human accidentally swallows eggs of *B. procyonis* (an infected raccoon can shed as many as 45 million eggs daily), the larvae emerging from these eggs migrate to organs as diverse as muscles, eyes, and brain, and it is not unusual for death to ensue. The first human case was detected in the United States in 1984 in an infant ten months old.

Human behavior is equally responsible for recent epidemics caused by prions (proteins with abnormal conformations). To begin with, new methods of feeding cattle caused the epidemic of mad cow disease in Great Britain, which peaked in 1993, and its appearance elsewhere. This disease probably appeared in cattle because they were fed sheep meat-and-bone meal that contained scrapie, and there is strong evidence that the outbreak was exacerbated by feeding rendered bovine meat-and-bone meal to young calves. The disease can then be transmitted to humans when they subsequently consume infected meat. Although diseases caused by prions had until then been sporadic and rare, the number of cases has suddenly jumped.

We should consider the group of diseases that preceded the original case of Lyme disease. This malady is caused by the bacterium *Borrelia burgdorferi*, which was discovered in 1977 in children in Connecticut. During 2002–3 it was responsible for more than 20,000 infections per year in the United States and has been found in many parts of the world. *B. burgdorferi* mainly infests deer and reaches humans through tick bites. Although the disease was identified only recently, examination of medical records shows that it existed at least by the early twentieth century, although the pathogen and means of infection were unknown. However, it is likely that increasingly frequent contacts between humans and nature (for example, through hiking) aided transfer.

Many examples similar to those I have just described can be cited.

When we speak of emerging diseases, hemorrhagic fever caused by the Ebola virus often comes to mind. However, it is quite possible that this disease is only a recently discovered one, like Lyme disease. Ebola virus causes brutal, fearsome epidemics in humans in central Africa, especially the Congo and Gabon. It is believed that, between epidemics, this virus persists in a "reservoir host" that has not yet been identified. We know only that there was at one time contamination of animals like gorillas, chimpanzees, and certain bovids (the ruminant family containing cattle, oxen, sheep, goats, and true antelopes) that came in contact with the reservoir host. It was only later that humans contracted the virus by contact with carcasses of these animals. The disease is severe in gorillas and chimpanzees—it can even endanger their populations (Leroy et al. 2004). Is it, in fact, a truly recent emerging disease? This is not certain because, given the epidemic nature of hemorrhagic fever, it is difficult to date the first human cases.

The Origin of Emerging Diseases

Emerging diseases seem, from an evolutionary standpoint, to be a price we pay for change, or even for progress. Behavioral changes have been numerous in the history and prehistory of humans, so it is not surprising that the emergence of new diseases should be a routine phenomenon. James Musser (1996, 15) wrote that "changes in human behavior, simple processes of microbial evolution, and increasing resistance to antimicrobial agents will continue to supply mankind with new infectious disease challenges." Now, at the dawn of the twenty-first century, humans have modified their behavior more than at any moment in their evolutionary or historical past, and it is probably not by chance that Musser placed behavior at the beginning of his warning. The fears associated with the use of pathogens in "bioterrorism" strikingly illustrate the importance of behavior.

Of course, when we speak of emerging diseases, we are mostly concerned with the most recent emergences. We think immediately of the drama produced by the AIDS pandemic. What is the origin of this virus, responsible for such a devastating illness? Scientists have thought at various times of a mutation of a virus that was originally "silent," of the simple spread by transport of a pathogen that was originally localized and thus passed unnoticed, of genetic recombination between strains that were individually not pathogenic and were placed in contact through mixing of populations, and finally of the passage to humans from another animal in Africa. It is this last hypothesis that is considered correct today. There are two distinct viruses, HIV1 and HIV2, each of which assumes several forms, identified by sequences in their genomes. These viruses infect various kinds of cells, but especially lymphocytes, the cells charged with immune defense. The virus HIV1 originated in chimpanzees and HIV2 in the sooty mangabey (a small, baboon-like African monkey). Thanks to molecular biological research, it is possible to go further back in the history of HIV1, which seems to have resulted from the fusion of genomes of several distinct viruses. These ancestral viruses, named SIV (for simian immunodeficiency virus), parasitize small monkeys and recombined in chimpanzees to yield the immediate ancestor of HIV1. From an ecological perspective, the most important question is how these viruses switched hosts. All the evidence suggests that the passage from small monkeys to chimpanzees occurred because the latter sometimes eat the former. Unfortunately, it appears that this explanation is equally valid for the passage from chimpanzees to humans. Bailes et al. (2003) write: "humans who hunt and handle bushmeat are exposed to a plethora of genetically divergent viruses." Humans have perhaps managed to become infected by cutting themselves in the course of butchering the remains of chimpanzees (for HIV1) and mangabeys (for HIV2). There are still poorly understood aspects of these transfers, especially the fact that the passage of the virus to the chimpanzee did not entrain a visible emergent disease (the lymphocytes, although infected, are not destroyed), whereas the passage to humans caused a virulent pandemic.

SARS (for severe acute respiratory syndrome), a pulmonary virus that appeared in China in 2003, is quite similar to AIDS in terms of the means of transfer to humans. In this case, it appears the virus arose in civets, small carnivores about the size of a large cat that are raised for food, especially in the Canton region.

A form of pneumonia caused by the bacterium *Legionella pneumophila*, Legionnaire's disease, is typically classed among emerging diseases. Its spread is tied to technical progress in that the pathogen is primarily found in aerosols produced by cooling towers.

West Nile virus is the causal agent of an encephalitis that qualifies as an emerging disease because it was unknown in humans before the 1930s. Transmitted to birds and mammals by mosquitoes during summer months, it reached the New World in 1999 and has already led to thousands of human cases. However, its status as a recently emerging disease is possibly belied by the suggestion that Alexander the Great might have died of West Nile fever (infection by the virus may be manifested by a fever, encephalitis, or both), in Baghdad, in 323 BC (Marr and Calisher 2002).

The appearance of hominids is a recent phenomenon in evolutionary time, so many transfers of pathogens are "young" or "recent" in this sense. When a disease passes to a new host, it can be particularly virulent simply because the immune system of the new host species is not ready to fight it. The classic example of the African trypanosomes illustrates this situation. Trypanosomes are unicellular parasitic animals; some species inhabit the blood of mammals. Their pathogenicity for African ungulates varies greatly depending on the origin of the ungulates. For wild ungulates, such as African antelopes and buffalo, trypanosome infections are endemic but not very pathological. For livestock imported to Africa several thousand years ago, the trypanosomes are more pathological and sometimes fatal, but it is nevertheless possible to raise these animals. Finally, for the European bovids that people have occasionally tried to establish in Africa recently, mortality has almost always been nearly 100 percent. In humans, trypanosomes cause African sleeping sickness, which is more virulent in eastern than in western Africa, leading to the hypothesis that this difference in pathogenicity is the result of the more recent acquisition of the pathogen in the East than in the West.

Questions about Emerging Diseases

The phenomenon of emerging diseases leads to many questions. One of them is, Do emerging diseases have a negative impact on human evolution?

When a transfer occurs, the newly afflicted human population is in fact in a situation similar to that of the Native Americans confronted with pathogens carried by the conquistadors. This leads to the question of how many millions of human deaths were needed from the dawn of humanity for the acquisition, surely by stages and surely gradual, of resistance to recently transferred parasites. We can also ask if such resistance was not a prerequisite for humans to be numerous enough to allow emigrations from the cradle of humanity in Africa. If this hypothesis is correct, it suggests that emerging diseases have served as a brake to human evolution. J. C. Eccles in 1989 emphasized that hominid evolution rested for a long time on tiny populations in constant danger of extinction. The human conquest of the temperate zones was an important step leading to the elimination of certain previously emerging diseases, because, in leaving the tropics, our ancestors probably reduced their burden of infectious pathogens. Even if infectious organisms were far from absent in temperate and cold regions, humans still could have benefited from the elimination of at least some of the diseases they had acquired in warmer areas; this loss is, in effect, a phenomenon that is the exact opposite of that of disease emergence. Mastery of fire, acquired some 400,000 years ago, could have lightened the load of parasitism because cooking destroys parasites transmitted to humans with food (such as trichinas, tapeworms, nematodes, and various flukes of the liver, intestines, and lungs).

A second question is, Have selective pressures generated by emerging diseases played any positive role in human evolution?

The most accurate response is that we have scant evidence to this effect. In fact, although defense against predators is one of the factors contributing to the selection of an ever-larger brain in hominids, the struggle against parasites has probably not played such a role. Paradoxically, the damage caused by parasites was not recognized until very recently (see Grove 1990): selective pressures caused by parasites may have resulted in the selection of genes coding for antibodies or other components of the immune system but probably were not a factor in the evolution of the synapses in our cerebral cortex. Even the knowledge of the existence of parasites as pathogenic agents is quite recent; perhaps it was the Egyptians (according to a papyrus of the eighteenth dynasty) or the Chinese (according to seventh-century manuscripts) who were the first to realize that schistosomiasis is caused by the penetration of the body by living organisms. And it was not until 1870 that Alphonse Laveran discovered minute *Plasmodium* individuals in the blood of malaria sufferers. So there has not been much time for human intelligence to contribute to the battle against parasitic diseases. We can only hypothesize that, since the role of parasites has been recognized, genes of individuals who were able to associate a particular behavior (such as eating a certain type of animal or bathing in certain types of water bodies) with a particular disease might have been selected for.

Finally, a third question is, Should we fear the emergence of new diseases in the future?

If we bear in mind the tendency for humans to engage in new behaviors, the risk of new diseases seems very real. As I have shown in a previous chapter that discusses research on ectoparasites of swifts, the parasites do not pay attention to the taxonomic position of their hosts. At least, they do not care *unless* the taxonomic position reflects the resources that the hosts

248   offer. Therefore, for a parasite to take advantage of the opportunity to acquire a new host, it is enough that humans constitute a new resource—so humans acquire a new parasite. Of course, it is impossible to make precise predictions about which parasites will transfer when. Nevertheless, it is possible to ask what the consequences might be of some global changes that humans are now inducing.

There have been enough indications that climate is presently warming at a significant rate to force us to consider the effect such change may have on human health. In various spots around the world, research (local, and occasionally conflicting) suggests that particular pathogens will emerge locally in response to ongoing climatic variation, such as the El Niño effect.

Unease is mainly about diseases that are vectored. Any change in the environment can have as a consequence the spread of a non-native vector into a new region or an increase in the capacity of a native vector to spread a pathogen (for instance, by an increase in population density, changes in the complex of competitors or other natural enemies, or acceleration of the life cycle). As I showed in chapter 3 with respect to malaria, an increase in transmission can lead to a greater diversity of parasite genotypes within afflicted individuals and therefore to increased virulence. However, the possibly relevant aspects of the biology of vectors are so numerous that the final effects of changes "currently remain nearly unpredictable," according to F. Rodhain (2002).

Worry is greatest where a disease has previously existed but has been substantially eliminated: can climatic changes cause it to reappear? For example, in Europe, inhabitants of Mediterranean regions still remember periods not so long ago when mosquitoes caused outbreaks of malaria, yellow fever, and dengue fever. Malaria, transmitted by mosquitoes in the genus *Anopheles,* was present in Europe until the beginning of the twentieth century; dengue fever, transmitted by mosquitoes in the genus *Aedes,* still caused major epidemics as late as 1920; and yellow fever, also transmitted by *Aedes,* ravaged Spain in the nineteenth century. The recent arrival in Europe of the Asian tiger mosquito, *Aedes albopictus* (carried in used tires), exacerbates the threat of a resurgence of the latter two diseases, just as its recent arrival in the United States has raised similar alarms.

Do these facts mean we must subscribe to the doomsday scenarios that predict the reemergence of these illnesses and others, such as tick-borne varieties of encephalitis? The response is probably no. In fact, as I emphasized earlier, many factors affect the transmission of vector-borne diseases. When we examine the matter closely, we see that, when malaria disappeared from Europe, climate probably played no role. In fact, in a way that can seem paradoxical, this disease was common in northern Europe during the period called the Little Ice Age during the sixteenth and seventeenth centuries. For example, it was present in London at a time when England was covered with snow several months each year. In other words, the disappearance of malaria from Europe did not depend on climate but on a complex of socioeconomic factors (for example, agricultural and sanitary practices and urbanization). Given these facts, and even though we should remain vigilant, it is hard to see how the greenhouse effect by itself could lead to a northern expansion of the range of malaria.

As these reflections indicate, the explosive metamorphosis of the hominid lineage and the changes that have ensued have not occurred without troublesome encounters with the living world already in place. Of course we can believe that parasites do not have a proper place in a terrestrial paradise. Unfortunately, they have always been present in the surrounding jungle, on the lookout for opportunity.

NOTES

Introduction

1. This is a translation by DS of Jacques Prévert's poem "Chanson des escargots qui vont à l'enterrement d'un feuille morte."

2. This division is called *cladogenesis*. It is often contrasted with *anagenesis,* which is the transformation of a species through time into a different species. This latter process is difficult to detect, and it does not change the number of branches on the tree.

3. This is almost always DNA (deoxyribonucleic acid), rarely RNA (ribonucleic acid). In more highly evolved species, DNA is organized into chromosomes, and each individual has pairs of homologous chromosomes, of which one member of each pair is of paternal origin and the other of maternal origin.

4. There are four types of nucleotides (A, T, G, C) in DNA, and each combination of three letters (for example, ATT) codes for a particular amino acid among the twenty that can be assembled into proteins. In RNA, T is replaced by U.

5. HIV stands for human immunodeficiency virus.

6. That is, about 10,000 billion parasitized red blood cells.

Chapter 1

1. Although I have stressed that it is the differential reproductive success of individuals that drives natural selection, I show in chapter 3 that it is ultimately differential reproduction of genes that is important. A gene can reproduce itself if it causes the individual bearing it to reproduce, but that is not the only way a gene can reproduce.

2. This divergence date is not universally accepted; some researchers believe the separation between chimpanzees and humans occurred much earlier.

3. This assertion is not always correct; for instance, some trypanosomes appear to provide vitamins (B6) to their hosts (Munger and Holmes 1988).

4. The sulfur-oxidizing bacteria are, for the biotic communities that inhabit the oceanic depths, the equivalent of green plants residing in communities that sunlight reaches. The other animals living in the vicinity of deep hydrothermal vents (crabs and fishes) are heterotrophic; that is, they must feed on other species, either the autotrophic bacteria themselves or the species that harbor these bacteria.

5. The proof is that the organs for capturing and digesting food in these bivalves have atrophied or even disappeared; the bivalves—that is, the hosts—have lost their digestive tract just as solitary worms (*Taenia*) have.

6. Sulfides are typically highly toxic molecules. Organisms that absorb sulfides to "feed" to their bacteria have biochemical adaptations (transport proteins) that allow them to take in the sulfides without damaging their tissues.

7. With respect to the idea of costs and benefits, nothing is truer than the following reflection of D. C. Smith and A. E. Douglas (1987): "an organism deriving benefit from an association need not do so from all the interactions with its partner; some interactions may be disadvantageous and represent a *cost*. The observed benefit is a *net benefit*, a balance between gross benefits and gross costs."

Chapter 2

1. To be precise, it is important to note that there are jellyfish (the narcomedusae) that parasitize other jellyfish, but this parasitism of one species in the phylum Coelenterata by another is but one case, compared with the parasitism that I will discuss. The ctenophores, a phylum closely related to the coelenterates, have no truly parasitic species.

2. According to E. V. Raikova, the cell with the small nucleus is the egg, whereas the cell with the large nucleus is the second polar body produced during meiosis.

3. Note that we have here an extraordinary situation, a highly organized, multicellular parasite parasitizing a single cell of its host.

4. This developmental process spread out over several years contrasts strikingly with that of free-living coelenterates, which takes at most two days.

5. There is no real justification for the sobriquet "solitary." The species of *Taenia* that parasitize humans, *T. solium* and *T. saginata,* are not always solitary.

6. This research entails sequencing the smaller subunit of the ribosomal transcription unit in 53 species of free-living and parasitic nematodes.

7. Throughout this work, I have used metaphors that should be interpreted as illustrations of natural selection and not as the result of any act of will or conscience on the part of the selected individuals.

8. This is perhaps the most beautiful example of a series of progressively adaptive stages to a new environment that is known in the entire field of evolution. The only other group that presents a comparable record is the copepods, of which about 500 species (of 1,500 total) have evolved to become parasitic.

9. In the development that I have traced, I have not accounted for all the species that might have been part of the series. Also, there is sometimes important variation among species of the same genus, which I have not taken into account. Finally, it is important to bear in mind that this depiction of parasitism by prosobranchs in the form of a series of stages is not meant to show the actual phyletic evolution of this taxon, which must have been "shrubby" rather than linear and tree-like. All the animals shown are extant, and therefore they cannot be ancestors of one another.

10. An exception would be in some associations in which the ants themselves excavate the domatia.

11. Unfortunately this type of research is in its infancy. To my knowledge, for example, the genome of *Enteroxenos* has not been compared with that of free-living prosobranchs.

12. In fact, Cope's law is not strictly observed in the family Equidae.

13. This diversity of hosts of this group, the Oxyuridae, for a single developmental stage (the adult) is unique among parasites.

14. Lateral transfer is a process to which I will often return. It consists of the passage of a parasite from one host lineage (in which the parasitism evolved) to a new host lineage not closely related to the original one.

15. This fact was first demonstrated for a transposable element called "gypsy" in *Drosophila* genomes when it was discovered that this transposable element could also act as a virus.

16. I should emphasize that, in this particular case, if there is a cost it is incurred by the individual, while any possible benefit accrues to the population.

17. This is an instance of a particular type of distortion of meiotic segregation, also known as meiotic drive. Normally, when there are two different alleles of a gene, any gamete formed at meiosis is equally likely to carry each of the alleles. An allele is said to be driven if it is capable of hindering the inclusion of the other allele in the gametes. The driven allele is therefore transmitted to more than half of the descendants, sometimes to all of them, and its frequency in the population grows until, in some cases, the other allele disappears completely. Details of the mechanism that produces meiotic drive are complex. The genome of *Rana ridibunda* acts as a driven genome.

Chapter 3

1. Only individuals in species capable of apparently endless division (such as unicellular species) and those that reproduce vegetatively (such as some plants and a few animals) can, in a sense, be considered immortal. An individual of a unicellular species dies only by accident and—in several species—by apoptosis (genetically programmed cell death).

2. This multiplication is generally ascribed to vegetative budding by Western researchers and to parthenogenesis by those of the East (Russia and Poland; see Dobrovolskij and Ataev 2003). In certain species, the succession of stages within the mollusc can be more complex than my description here.

3. Not all species of *Microphallus* have a shortened life cycle. We will soon see that the species *M. papillorobustus* has a classic three-host cycle.

4. A question arises naturally: how was the three- host life cycle considered typical acquired in the course of evolution? This question is not yet resolved. Some authors believe trematodes were originally parasites of molluscs and that the cycle was subsequently lengthened. Others think vertebrates were the original hosts and that the larval forms secondarily became parasites of molluscs, then of the second host.

5. Some selective processes can be quick (the time scale can be several weeks or even several days in microparasites for traits associated with virulence), but the multiple changes required for the evolution of a life cycle can probably be produced only on scales of several million years.

6. In this research, individuals are classified as having different genotypes if the method of DNA analysis used, known as *random amplification of polymorphic DNA,* shows differences in the nucleotide sequence.

7. Here I am employing an elementary application of C. E. Shannon's information theory.

8. Ectoparasites are subject to the direct influence of the external environment; endoparasites are separated from the external environment by host tissues; mesoparasites are located in host cavities (such as the digestive tract) that communicate with the exterior (Euzet 1989).

9. It is possible that some insect parasitoids use sight to identify future victims, but even if they do, other signals (especially odors and vibrations) are always the main ones. G. Breton informs me that the large eyes of *Anilocra* (isopod parasites of fishes) may be used to detect hosts.

10. In chemical ecology, we distinguish (*a*) molecules that signal intraspecifically, or pheromones; (*b*) molecules that signal interspecifically and benefit the emitting species, or allomones; and (*c*) molecules that signal interspecifically and benefit the receiving species, or kairomones. Signals emitted by hosts and detected by infective stages of parasites are therefore kairomones.

11. The ovipositor of the ichneumonid wasp *Megarhyssa atrata* is up to 14 centimeters long (Nenon, Kacem, and Le Lannic 1997). It can thus penetrate 14 centimeters of wood to deposit its eggs on larvae of another hymenopteran, *Tremex columba.*

12. F. Athias-Binche informs me that this mite can jump up to 5 centimeters. In light of its size, this would be the equivalent of a human's jumping 500 meters high.

13. The host a parasite leaves is the upstream host; the host a parasite goes to is the downstream host (Combes 1995). In a holoxenic life cycle (during which there is one host), there is no reason why the upstream and downstream hosts cannot belong to the same species. In a heteroxenic cycle (during which there are several hosts), upstream and downstream hosts nearly always belong to different species.

14. In this regard, I should add that woodcocks have another reason for appearing in a book devoted to associations between living organisms. In fact, every woodcock harbors millions of small tapeworms of the genus *Amoebotaenia* in its digestive tract. Now, all gourmets know that woodcocks are consumed with their intestines intact. The reason is that it is precisely these tapeworms that give the woodcock its delicious taste, or at least contribute strongly to it (René Grau, pers. comm.).

15. Although the distance between the emitter and the target is only a few dozen centimeters in lagoons, it can exceed 10 meters on marine coasts.

16. Or the eggs may cross the wall of the bladder in the case of urinary bilharziasis.

17. In these two bivalves, a small fraction of the metacercariae also settle in the general extrapallial space, just beneath the hinge.

18. The word *invasion* is not an exaggeration; there can be up to 5,000 metacercariae per cockle and 10,000 per clam.

19. I should emphasize that P. Bartoli has conducted all the experiments required to prove that this is not a case of two parasite species being mistakenly lumped together.

20. In some crustacean populations, males can become very rare. They never disappear entirely, however, because the increasing rarity of males selects, in the host genome, for "masculizing" genes that counter the feminization caused by *Wolbachia*.

21. This association illustrates the concept of "parasitizing a near relative" presented in the introduction.

22. Recall that bees, wasps, and ants are hymenopterans, whereas termites belong to a different insect order, the Isoptera.

23. The cited authors mainly apply the concept of "constructing a niche" to the human species, in which cultural activities greatly increase the capacity to modify selective pressures imposed by the environment. Control of infective agents and parasites is one aspect of the cultural construction of the niche.

24. This pathogen might be a virus, bacteria, protist, fungus, or metazoan, and not only a virus, as might be imagined from the root of the word.

25. Of course, one must bear in mind that optimal virulence is optimal only for the pathogen. For the host, the only optimal virulence is zero virulence.

26. Because it is shaped differently in different species, the spur is useful for discriminating easily among human parasites during diagnostic examinations. For example, *Schistosoma mansoni* eggs have a highly characteristic lateral, slightly recurved spur.

27. In fact, all eggs do not pass to the exterior, even when the schistosomes encounter "normal" conditions. Eggs are laid in fine veinules that surround the digestive tract, encountering the blood flowing from the intestine to the liver. Some eggs are therefore entrained toward the liver and form granulomas there. This drift of some eggs to the liver has two consequences. One is that these eggs are lost to the life cycle, because they can never exit to the exterior. The other is that these eggs are the main pathological agent in humans, because the granulomas will eventually be replaced by scar tissue that cannot perform liver functions.

28. The name *Dracunculus* means "little dragon" and was given by Plutarch in the first century AD; *medinensis* refers to Medina, named by the Muslim physician Avicenna in the eleventh century.

29. Bernardin de Saint-Pierre was a French author and naturalist (1737–1814) whose "natural theology" argued that everything in nature was designed and well suited for some purpose.

30. Gould and Lewontin were not critical of Bernardin de Saint-Pierre but rather of the application of adaptationist ideas propounded in the 1970s by some sociobiologists about human behavior.

31. At the top of each spandrel is one of the four Evangelists. At each base is one of the biblical rivers: the Tigris, the Euphrates, the Nile, and the Indus.

32. Even if this demographic aspect is not evident and, in any case, has not been demonstrated, Louis Euzet has pointed out to me that there is perhaps a trait in the ancestors of *Helipegus* that predisposes the genus in this direction. In fact, in many species in the Hemiuridae, a "waiting host" (or paratenic host) can be inserted between the copepod and the definitive host. This paratenic host is usually a small fish, and the

256 metacercariae survive in its intestine. This is exactly what happens in the dragonfly, and this stage has become obligatory.

33. Janice Moore (1993, 124) wrote, "such an animal is worthy of our liveliest curiosity, if not our frank admiration."

34. The comparative aspect of this phrase refers to the heated controversy, sometimes exaggerated in one direction, sometimes in the other, following the publication of the article by Gould and Lewontin.

35. I am grateful to Bernard Chaubet, who provided copious documentation of the biology of this species.

36. An interesting detail is that fishes parasitized by *E. californiensis* also harbor metacercariae of another trematode, *Renicola buchanani,* but these are always located in the liver. Because this trematode matures in the same birds as *E. californiensis* and it is almost certainly the latter species that causes the modified behavior of the fishes, *R. buchanani* has its transmission improved without itself investing in the favorization process. A similar example of favorization piracy was discovered by Simone Helluy in 1982 in the Camargue: the trematode *Microphallus papillorobustus,* whose metacercariae lodge in certain nerve centers in *Gammarus* amphipods (see earlier paragraphs), entrains in its wake the transmission of a trematode that does not settle in any particular site, *Maritrema subdolum.*

Chapter 4

1. Of course, the choice exists between different methods of avoiding the encounter (camouflage, freezing, detecting predators, flight, etc.). By the word *choice,* in this chapter, I mean the different options that natural selection can produce.

2. The great tit and the blue tit can compete for food, as witnessed by the fact that clutch size and weight of the young of the great tit are lower in forests with the highest blue tit densities. It appears that the blue tit is the principle "source" of the flea *Ceratophyllus gallinae,* which attacks both tit species. Exchange of fleas may occur through exchange of nests. It is therefore possible that the diminution in reproductive success of great tits where blue tits are abundant is related to the increase in flea density, with the fleas in a sense acting as allies of the blue tits (Richner, Oppliger, and Christe 1993).

3. This book is not an immunology text. That is why I discuss the relationship between immunity and different aspects of the lives of hosts without analyzing the mechanism of immunity itself.

4. The word *richness,* derived from the concept of species richness of ecosystems, is obviously jarring in the context of parasitism, but it is consecrated by traditional usage.

5. Of course, it is necessary to bear in mind that these categories are arbitrary, because if we go back far enough in time, we always end up encountering a transfer, even if it is from the abiotic environment (passage from a free-living condition to parasitism).

6. This is true even though the pathogen of mad cow disease (in humans, Creutzfeldt-Jakob syndrome) is a very special infective agent, because it is an abnor-

mally shaped protein that, surprisingly, can transmit its anomaly to normal protein molecules.

7. *S. haematobium* is truly restricted to humans; *S. mansoni* uses humans as its main host but can, in certain settings, also infect rodents, a trait that is a sort of "souvenir" of its ancestor, a parasite of murid rodents.

8. For *Schistosoma* as for *Onchocerca*, determination of phyletic parentage rests on DNA analysis.

Chapter 5

1. I use the term *mutualist* more than *symbiosis* because it more clearly indicates that the association yields benefits to both partners. The word *symbiosis*, although often construed as having the same sense, from an etymological perspective denotes only a "life together," whether of a parasitic or mutualistic nature. At the end of the nineteenth century, the originator of the term *symbiosis*, the German du Bary, included in this designation all forms of association between two individuals of different species.

2. Sea anemones are coelenterates with tentacles bearing urticating cells, and not plants, as their name might suggest.

3. Note that I retain the quotation marks around *domesticators*.

4. Another category of virus, called ascovirus, plays a similar role, but it multiplies in host caterpillars, which it can destroy, to its own benefit. These viruses can be considered as parasites of the caterpillars, and they use parasitoids as vectors (just as the *Plasmodium* that causes malaria uses mosquitoes to get itself injected into humans).

5. In general, certain lines have been recognized as pathogenic if ingested (see Bowen et al. 1998).

6. Species of *Photorhabdus* have five genes, named *luxA* through *luxE*, that are required for light production.

7. This simplified description of the life cycle does not take sexual reproduction into account. Many other pathogen life cycles resemble that of *Plasmodium;* for example, fleas transmit the plague bacterium *Yersinia pestis* among rats and humans; a single injected bacterial cell can be infectious and lethal.

8. The infrapopulation is the ensemble of all parasites of a given species present at a particular moment in the same host individual.

9. These antibacterial peptides are defensive molecules typically present in invertebrates (for example, the cecropins).

10. In particular, the bacteria inhibit phenoloxidase, an essential enzyme in the cascade of reactions, leading to encapsulation and melanization of foreign bodies by insects.

11. Boemare et al. (1997, 35) write: "Nematodes need a special 'menu,' and the most suitable for their reproduction is a medium prepared by their symbionts and the bacterial biomass too."

12. This perfect quiescence of the bacteria in the digestive tract of the nematode has yet to be explained. Similarly unexplained is the fact that the nematode first consumes

bacteria but finally reserves a "sanctuary" for them in their gut. As Forst et al. (1997, 68) write, "How do the bacteria and nematode communicate?"

13. There are fig species in New Guinea and New Caledonia that have conserved the ancestral trait of a fig in the shape of an open urn.

14. In most fig species, the ostiole is so constricted that the agaonids lose their wings passing through it and therefore generally never leave a fig they enter. In certain wasp–fig associations, occasional wasps may go to another fig on the same tree (Giberneau et al. 1996; Moore et al. 2003).

15. I have simplified certain aspects of the fig–agaonid relationship. The main two complications are (*a*) some figs are dioecious (certain trees have male fruits, whereas others have female fruits); (*b*) the contrast between the long larval life and the short life of the adult wasps mandates that there will be figs at different stages of maturity at the same place and time. For more detailed knowledge about these matters, see the many publications by Finn Kjellberg and his colleagues.

16. It is evident that, in the cleaner fish–client relationship, each interaction does not last long.

17. Such a role as carrier has also been suggested for mutualistic fungi of the genus *Armillaria* in the association between the angiosperm *Monotropa hypopitys* and diverse species of gymnosperms.

18. Carnivorous plants generally have poorly developed roots, which proves that the garnering of organic matter owing to the capture of insects is of more than anecdotal importance.

19. That is why we provide plants with fertilizer in the form of nitrates, excessive use of which can pollute groundwater.

20. I am borrowing some of the data from a paragraph that follows the excellent explanation of J. Denarié, published on the occasion of the lecture series The University of All Knowledge, in 2000.

21. Young and Johnston (1989) write that, in some cases, species of *Rhizobium* "are better taxonomists than the taxonomists themselves."

22. The rhizosphere is the immediate environment of the roots.

23. With respect to the association of ants and plants, we are driven to ask, How can such plants be pollinated if the ants prevent other insects from approaching? The solution produced by natural selection is ingenious. Willmer and Stone (1997) have shown that, at the exact moment when its flowers should receive insect pollinators, the plant emits a volatile chemical signal that momentarily repels ants, and only ants, from the flowers. After fertilization, this signal is no longer emitted and the ants resume patrolling, thus protecting the fertilized ovules and developing seeds.

24. We distinguish two kinds of living cells, the prokaryotic cell and the eukaryotic cell. The prokaryotic cell is the bacterial cell. It comprises a plasma membrane that is generally paralleled toward the outside by a rigid wall, cytoplasm, and free DNA in the cytoplasm in the form of a single chromosome, which is eventually accompanied by small supplementary circles of DNA, the plasmids. The eukaryotic cell is that of all species located "above" bacteria in the tree of life. It includes, as does the prokaryotic

cell, a plasma cell membrane (protected by a cell wall in plants) and cytoplasm. In the cytoplasm, it contains a nucleus bounded by a double membrane pierced by pores. The nucleus harbors the DNA in the form of chromosomes, the number of which varies with the species. Whether the DNA is free or enclosed in a nucleus is not the only way in which prokaryote and eukaryote cells differ. The most important difference is that the cytoplasm of the eukaryote cell typically contains different organelles, among which are mitochondria and chloroplasts (the latter only in plant cells).

25. Humans are constructed of about a million billion cells.

26. Leigh and Rowell (1995) compare mitochondria to farmers who, because they have no new land to cultivate, have no solution for survival other than to take care of the land they already own.

27. I do not discuss the chloroplast case, which is more complicated.

28. In thyme, for example, the proportion of individuals that are purely female can reach 90 percent (Gouyon, Henry, and Arnould 1997).

29. Each gene is situated at a precise spot in the DNA molecule, its locus. In diploid organisms (with $2n$ chromosomes), there are thus two genes at each locus. Alleles are different forms that can be taken by the gene at a particularly locus. If the two alleles are identical, the individual is said to be homozygous at that locus; if they are different, it is heterozygous.

Chapter 6

1. A concrete illustration of the selection process is provided by "white lions," which reproduce perfectly well in zoos but are selected against in nature because ungulates can spot them from far away.

2. For detailed biology of the cuckoo, see two remarkably illustrated numbers (38 and 39) of the superb French children's nature magazine *La Hulotte* (08240 Boult-aux-Bois, France).

3. Unlike mammals, in which females have two X chromosomes and males one X and one Y, in birds, males have two M chromosomes, and females have one M and one N.

4. The most meaningful image of this process is to imagine Alice and the Queen walking on an airport moving sidewalk against the direction of motion.

5. One of the authors of the neo-Darwinian synthesis.

Chapter 8

1. Some ecologists see little value in comparisons of abundance among species, because comparisons among populations within the same species show major differences. Arneberg, Skorping, and Read (1997) discuss this problem.

2. There are also intermediate levels of analysis, such as human families or bird clutches (see Boulinier, Ives, and Danchin 1996).

3. To my knowledge, the existence of 17 species of the same genus on one host cannot be far from a world record.

4. I am speaking here of insectivorous birds such as blackbirds for *L. paradoxum* and shorebirds such as gulls for *M. pygmaeus*.

5. Monogeneans are flatworms related to trematodes and tapeworms, but their life cycle is direct (with no intermediate host). Nearly all are ectoparasites, living on the skin and gills of fishes. However, one family—the Polystomatidae—have conquered the nonfish world and become mesoparasites. They are found on amphibians, including the giant Japanese salamander, and freshwater turtles, and one species even lives under the eyelids of hippopotami.

6. I am not counting other, rarer species in Asia and Africa that also affect humans.

7. If the proposed phylogenetic tree is correct, it implies that the species *turkestanicum* should be transferred to the genus *Schistosoma*.

8. *P. integerrimum* parasitizes the common frog *Rana temporaria* in Europe; *P. americanus* parasitizes the desert toad *Scaphiopus couchii* in Arizona. The larval form of *P. integerrimum*, which attaches to the gills of tadpoles, can also, if the tadpole is very young, mature in place. In this case, it dies at metamorphosis. All these members of the Polystomatidae are related to *Eupolystoma alluaudi*, which I cited at the beginning of this chapter with respect to its reproduction in the bladder of an African toad.

9. Louis Euzet (1956) and Caira, Jensen, and Healy (2001) described, in the group of tetraphyllid tapeworms that parasitize the spiral valvule of sharks and rays, a series of scolex forms each more extravagant than the last.

10. The title of this section is borrowed from a work by Alex Kahn (2000).

11. The distance to water from the living place and also behavior with respect to water always introduce disparities in the distribution of water-dependent parasites of humans: bathing frequency, often related to age, profession, sex, and traditions, causes certain individuals to be much more exposed than others (an individual who never washes would never contract a disease like schistosomiasis).

12. This observation with respect to human habitations calls to mind the study by Thierry Boulinier, Ives, and Danchin (1996), which shows that the unit of aggregation of the tick *Ixodes uriae* in chicks of the gull *Rissa tridactyla* is the nest and not the individual (differences are significant between clutches, not between chicks of the same clutch).

13. The authors note, however, that these practices and behaviors were not elaborated to prevent parasitic diseases but rather for completely different reasons. We can say these are "cultural exaptations," to transfer to a different context the concept of exaptation proposed by Gould and Vrba (1982).

Chapter 9

1. The reader will have observed that throughout this book I have not hesitated to name as living beings not only viruses but also the self-replicating molecules (DNA and RNA) themselves. Obviously, this viewpoint can be debated, especially because a virus or a molecule does not "live" if it is isolated. In fact, it all depends on how we define life. For me, life is the ability to re-create a complex structure from elements garnered from the environment.

2. If the changes are rapid and occur in populations reduced in size, this would explain why paleontologists rarely find fossils of forms that are intermediate between

species of very distinct appearance. This hypothesis predicts that forms undergoing rapid evolution are local and temporary, thus rare, and therefore their fossil remains are unlikely to be found, a prediction that corresponds more to observations than that implied by a gradualist hypothesis.

3. Conversely, analyses of DNA sequences have shown that, among the ensemble of point mutations that arise, only a small number have an effective role in evolution.

4. I chose this example because such adaptations lead us to think, with Louis Euzet, that tapeworms may be the oldest metazoan parasites and have existed for hundreds of millions of years.

5. More precisely, there are several precedents. For example, bird songs are transmitted partly in Lamarckian fashion.

6. Taddei, Matic, and Radman (2000) recall that Lewis Carroll's Red Queen said not only that to stay in the same place, one must run very fast, but also that "if you want to get somewhere else, you must run at least twice as fast as that!"

7. The so-called resistant strains of *Plasmodium falciparum,* the agent of malaria, are spreading further and further. William D. Hamilton, one of the great evolutionists of our age, whom I have mentioned several times in this book, died in February 2000 from complications of malaria contracted in Africa and resistant to all treatments.

GLOSSARY

actin: One of the two proteins involved in muscle contraction

annelid: Member of the phylum Annelida, which includes earthworms, many marine worms, and leeches

antigen: Foreign substance that triggers the production of antibodies in an immune response

autotroph: Organism that synthesizes its own food from simple substances

barbule: One of the processes along the edge of the barb of a feather

bovid: Mammal species belonging to the family Bovidae, including cattle

buccal cavity: Cavity inside the mouth, through which food is ingested

centriole: Short cylindrical organelle, found in pairs in most cells of animals, fungi, and algae

centromere: Condensed region of chromosome that appears during mitosis where the chromatids are held together

chloroplast: Organelle in plant cells and some protists that contains chlorophyll and is the site of photosynthesis

chromatography: Separation of chemical substances by use of differences in the rates at which they travel through or along a stationary medium

codon: Three bases (nucleotides) in a DNA or RNA sequence that specify an amino acid or a termination signal (stop codon)

complement: One of a series of proteins in the blood serum that are part of an immune response

conspecific: Belonging to the same species

demographic: Characteristic of a population, such as growth rate, size, or sex ratio

eukaryote: Cell or organism with membrane-bound, structurally discrete nucleus and other subcellular compartments. Eukaryotes include all organisms except viruses, bacteria, and blue-green algae.

exaptation: Use of a structure or feature for a function other than that for which it evolved through natural selection

gamete: Mature male or female reproductive cell (sperm or ovum) with a haploid set of chromosomes

genome: All the genes of a particular organism

genotype: All the genes of a particular organism, as opposed to its observable traits—the phenotype

hemocoel: A cavity or series of spaces between the organs of most arthropods and molluscs through which the blood circulates

heteroxenic life cycle: Parasite life cycle that includes more than one host species in a sequence

holoxenic life cycle: Parasite life cycle that includes only one host species

homeotic gene: Genes in segmented organisms that specify the identity of each body segment by controlling the identity of the organs that develop within that segment

horizontal transmission: Transmission of parasites between host individuals of the same generation

infrapopulation: All the individuals of one parasite species present in one host individual

locus: Location of a gene on a chromosome

lymphocyte: Type of white blood cell produced in the lymphoid organs that is mainly responsible for immune responses. Present in the blood, lymph, and lymphoid tissues.

meiosis: Process consisting of two consecutive cell divisions in the diploid progenitors of sex cells, producing four daughter cells, each with a haploid set of chromosomes

Metazoa: All multicellular animals except sponges. Does not include protozoans.

mitochondrion: Organelle in cells of most eukaryotes, site of most energy production in most eukaryotes

mitotic spindle: Microtubule-based structure present during mitosis (cell division) to which chromosomes attach, after which they move toward opposite sides of the dividing cell

mycorrhizae: Hyphae of certain fungi (mycorrhizal fungi) that are symbiotic with plant roots on which they occur. Also, the combination of these hyphae and the plant roots.

myosin: Contractile protein that interacts with the protein actin to cause muscle contraction or cell movement

nucleotide: Building block of DNA and RNA molecules, consisting of a nitrogenous base, a five-carbon sugar, and a phosphate group. Groups of three nucleotides form codons, which are strung together to form genes, which in turn are strung together to form chromosomes.

ocellus: Simple eye of some insects, sensitive to light but incapable of forming visual images. Also, an eye-like marking, such as on some butterfly wings.

oocyte: Female gametocyte (sex cell) that develops into an ovum after undergoing meiosis

osmotrophy: Nutritional mode in which an organism acquires food by actively transporting it through the cell membrane in a dissolved state

oviposit: To lay or deposit eggs

parasitoid: Parasitic insect, especially a wasp or fly, that completes its growth inside a single host, eventually killing it

parthenogenesis: Development of an adult from an unfertilized egg

passerines: Perching birds in the order Passeriformes, mostly small songbirds

peroxisome: Organelle in the cytoplasm of eukaryotic cells, containing enzymes used in energy production

phagocytosis: Process by which phagocytes (such as white blood cells) surround and destroy microorganisms or other foreign matter

phenotype: The observable traits of an organism as opposed to the set of genes it possesses (its genotype)

phytophagous: Feeding on plants

plasmid: Autonomously replicating extrachromosomal circular piece of DNA, usually referred to in bacteria but distinct from the normal bacterial genome

polymorphism: Having more than one form; specifically, the existence in substantial frequencies of more than one genotype in a population

prokaryote: Cell or organism (such as a bacterium) lacking a membrane-bound, structurally discrete nucleus and other subcellular compartments

propagule: Dispersal stage of a plant or animal, such as fertilized eggs, larvae, or seeds. Also, individual or individuals of any organism that found a population.

proteolytic enzyme: Enzyme that catalyzes the splitting of proteins into smaller peptides and amino acids

Protista: Eukaryotic one-celled living organisms distinct from multicellular plants and animals, including protozoa, slime molds, and eukaryotic algae

retrovirus: Virus (such as HIV) with genetic material in the form of RNA rather than DNA that uses the enzyme reverse transcriptase to transcribe it into DNA

rhizosphere: Area of soil near plant roots, often containing large populations of microorganisms

ribosome: Subcellular particle, composed of RNA and protein, that participates in synthesis of protein in the cell

scolex: Knoblike anterior end of a tapeworm, with suckers or hooks that attach the tapeworm to its host

sequence homology: Degree of similarity between sequences, especially of nucleotides in DNA or RNA and amino acids in proteins

sexual dimorphism: Difference (usually morphological) between males and females of the same species

sympatric speciation: Speciation without geographic isolation

tarsus: Insect foot, consisting of from one to five segments attached to the tibia. Also, part of the leg between knee and foot.

taxon: Group of organisms of any taxonomic rank, such as a family, genus, or species

telomere: Terminal part of a eukaryotic chromosome, consisting of a few hundred base pairs with a defined structure

urticating: Stinging, as by some insect hairs

vertical transmission: Transmission of parasites from parent to offspring

REFERENCES

Abel, L., F. Demenais, A. Prata, A. E. Souza, and A. Dessein. 1991. Evidence for the segregation of a major gene in human susceptibility/resistance to infection by *Schistosoma mansoni*. *American Journal of Human Genetics* 48:959–70.

Alvarez, F. 1993. Proximity of trees facilitates parasitism by cuckoos *Cuculus canorus* on rufous warblers *Cercotrichas galactotes*. *Ibis* 135:331.

Anderson, R. M., and R. M. May. 1982. Coevolution of hosts and parasites. *Parasitology* 85:411–26.

Anstett, M. C., M. Hossaert-McKey, and F. Kjellberg. 1997. Figs and fig pollinators: Evolutionary conflicts in a coevolved mutualism. *Trends in Ecology and Evolution* 12:94–99.

Arnal, C., and S. Morand. 2001a. Cleaning behaviour in the fish: *Symphodius melanocercus*. Females are more honest than males. *Journal of the Marine Biological Association of the United Kingdom* 81:317–23.

———. 2001b. Importance of ectoparasites and mucus in cleaning interactions in the Mediterranean sea. *Marine Biology* 138:777–84.

Arneberg, P., A. Skorping, and A. F. Read. 1997. Is population density a species character? Comparative analysis of the nematode parasites of mammals. *Oikos* 80: 289–300.

Aron, S., and L. Passera. 2000. *Les sociétés animales: Evolution de la coopération et organisation sociale*. Paris: De Boeck Université.

Athias-Binche, F. 1990. Sur le concept de symbiose: L'exemple de la phorésie chez les acariens et son évolution vers le parasitisme ou le mutualisme. *Bulletin de la Société Zoologique de France* 115:77–98.

———. 1994. *La phorésie chez les acariens: Aspects adaptatifs et évolutifs*. Perpignan, France: Editions du Castillet.

Athias-Binche, F., and S. Morand. 1993. From phoresy to parasitism: The example of mites and nematodes. *Research and Reviews in Parasitology* 53:73–79.

Aubry, P. 1979. L'expédition française de Madagascar de 1895. Un désastre sanitaire. Pourquoi? *Médecine et Armées* 7:745–52.

Bagnères, A.-G., M. C. Lorenzi, G. Dusticier, S. Turillazi, and J.-L. Clément. 1996. Chemical usurpation of a nest by paper wasp parasites. *Science* 272:889–92.

Bailes, E., F. Bibollet-Ruche, V. Courgnaud, M. Peeters, P. A. Marx, B. Hahn, and P. M. Sharp. 2003. Hybrid origin of SIV in chimpanzees. *Science* 300:1713.

267

268

Bakke, T. A. 1980. A revision of the family Leucochlorchiidae Poche (Digenea) and studies on the morphology of *Leucochloridium paradoxum* Carus, 1835. *Systematic Parasitology* 1 (3–4): 189–202.

Barbault, R. 1994. *Des baleines, des bactéries, et des hommes.* Paris: Odile Jacob.

Bartoli, P. 1973a. La pénétration et l'installation des cercaires de *Gymnophallus fossarum* P. Bartoli, 1965 (Digenea, Gymnophallidae) chez *Cardium glaucum* Bruguière. *Bulletin du Muséum National d'Histoire Naturelle* 117, *Zoologie* 91:319–33.

———. 1973b. Les microbiotopes occupés par les métacercaires de *Gymnophallus fossarum* P. Bartoli, 1965 (Trematoda, Gymnophallidae) chez *Tapes decussatus* L. *Bulletin du Muséum National d'Histoire Naturelle* 117, *Zoologie* 91:335–48.

———. 1978. Modification de la croissance et du comportement de *Venerupis aurea* parasité par *Gymnophallus fossarum* P. Bartoli, 1965 (Trematoda, Digenea). *Haliotis* 7:23–28.

———. 1989. Les trématodes digénétiques parasites marqueurs de la biologie des goélands leucophées *Larus cachinnans michaellis* en Corse (Méditerranée occidentale). *Vie Marine* 10:17–26.

Bartoli, P., and C.-F. Boudouresque. 1997. Transmission failure of parasites (Digenea) in sites colonized by the recently introduced invasive alga *Caulerpa taxifolia*. *Marine Ecology Progress Series* 154:253–60.

Bartoli, P., M. Bourgeay-Causse, and C. Combes. 1997. Parasite transmission via a vitamin supplement. *BioScience* 47:251–53.

Bartoli, P., and C. Combes. 1986. Stratégies de dissémination des cercaires de trématodes dans un écosystème marin littoral. *Acta Oecologica Oecologia Generalis* 7: 101–14.

Barton, N. H. 2001. Speciation. *Trends in Ecology and Evolution* 16:325.

Batra, L. R., and S. W. T. Batra. 1985. Floral mimicry induced by mummy-berry fungus exploits host's pollinators as vectors. *Science* 228:1011–13.

Beehler, B. 1990. Les oiseaux de Paradis. *Pour la Science* 148:38–45.

Bentz, S., S. Leroy, L. Du Preez, C. Vaucher, J. Mariaux, and O. Verneau. 2001. Origin and evolution of African *Polystoma* (Monogenea: Polystomatidae) assessed by molecular methods. *International Journal for Parasitology* 7:697–705.

Bergstrom, C. T., and M. Lachmann. 2003. The Red King effect: When the slowest runner wins the coevolutionary race. *Proceedings of the National Academy of Sciences* (USA) 100:593–98.

Berticat, C., F. Rousset, M. Raymond, A. Berthomieu, and M. Weill. 2002. High *Wolbachia* density in insecticide-resistant mosquitoes. *Proceedings of the Royal Society of London B* 269:13–16.

Black, F. L. 1992. Why did they die? *Science* 258:1739–40.

Blaxter, M. L., P. De Ley, J. R. Garey, L. X. Liu, P. Scheldeman, A. Vierstraete, J. R. Vanfleteren, L. Y. Mackey, M. Dorris, L. M. Frisse, J. T. Vida, and W. K. Thomas. 1998. A molecular evolutionary framework for the phylum Nematoda. *Nature* 392:71–75.

Boemare, N., A. Givaudan, M. Brehelin, and C. Laumond C. 1997. Symbiosis and pathogenicity of nematode–bacterium complexes. *Symbiosis* 22:21–45.

Bonavita-Cougourdan, A., A. G. Bagnères, E. Provost, G. Dusticier, and J. L. Clément. 1997. Plasticity of the cuticular hydrocarbon profile of the slave-making ant *Polyergus rufescens* depending on the social environment. *Comparative Biochemistry and Physiology B* 116:287–302.

Bond, W. J. 1989. The tortoise and the hare: Ecology of angiosperm dominance and gymnosperm persistence. *Biological Journal of the Linnean Society* 36:227–49.

Bouix-Busson, D., D. Rondelaud, and C. Combes. 1985. L'infestation de *Lymnaea glabra* Müller par *Fasciola hepatica* L. *Annales de Parasitologie Humaine et Comparée* 60: 11–21.

Boulinier, T., A. R. Ives, and E. Danchin. 1996. Measuring aggregation of parasites at different host population levels. *Parasitology* 112:581–87.

Bowen, D., T. A. Rocheleau, M. Blackburn, O. Andreev, E. Golubeva, R. Bhartia, and R. H. French-Constant. 1998. Insecticidal toxins from the bacterium *Photorhabdus luminescens*. *Science* 280:2129–32.

Bowles, S., and H. Gintis. 2002. Homo reciprocans. *Nature* 415:125–28.

Brillard, J., C. Ribeiro, N. Boemare, M. Brehélin, and A. Givaudan. 2001. Two distinct hemolytic activities in *Xenorhabdus nematophila* are active against immunocompetent insect cells. *Applied and Environmental Microbiology* 67: 2515–25.

Caira, J. N., K. Jensen, and C. J. Healy. 2001. Interrelationships among tetraphyllidean and lecanicephalidean cestodes. In *Interrelationships of the Platyhelminthes*, ed. D. T. J. Littlewood and R. A. Bray, 135–58. London: Taylor and Francis.

Caro, A., C. Combes, and L. Euzet. 1997. What makes a fish a suitable host for Monogenea in the Mediterranean? *Journal of Helminthology* 71:203–10.

Caullery, M. 1950. *Le parasitisme et la symbiose*. Paris: G. Doin and Cie.

Chaubet, B. 1996. Le patient parasite de la mante religieuse. *Pour la Science* 219:106.

Chen, L., K. V. N. Rao, Y.-X. He, and K. Ramaswamy. 2002. Skin-stage schistosomula of *Schistosoma mansoni* produce an apoptosis-inducing factor that can cause apoptosis of T-cells. *Journal of Biological Chemistry* 277:34329–35.

Cheng, T. C. 1970. *Symbiosis, organisms living together*. New York: Pegasus.

Cherrier, J. F., M. Gondran, P. Woltz, and G. Vogt. 1992. Parasitisme interspécifique chez les Gymnospermes: Données inédites chez deux Podocarpaceae endémiques néo-calédoniennes. *Revue de Cytologie et Biologie Végérale, Botanique* 15:65–87.

Christe, P., H. Richner, and A. Oppliger. 1996. Begging, food provisioning, and nestling competition in great tit broods infested with ectoparasites. *Behavioral Ecology* 7: 127–31.

Clay, T. 1957. The Mallophaga of birds. In *Premier symposium sur la spécificité parasitaire des parasites de vertébrés*, 120–58. Neuchâtel, Switzerland: Imprimerie Paul Attinger.

Clayton, D. H. 1990. Mate choice in experimentally parasitized rock doves: Lousy males lose. *American Zoologist* 30:251–62.

Cockburn, T. A. 1971. Infectious diseases in ancient populations. *Current Anthropology* 12:45–62.

270   Coluzzi, M. 1999. The clay feet of the malaria giant and its African roots: Hypotheses and inferences about origin, spread and control of *Plasmodium falciparum*. *Parasitologia* 41:277–83.

Combes, C. 1980. Les mécanismes de recrutement chez les métazoaires parasites et leur interprétation en termes de stratégies démographiques. *Vie et Milieu* 30:55–63.

———. 1991a. Ethological aspects of parasite transmission. *American Naturalist* 138:866–80.

———. 1991b. Where do human schistosomes come from? *Trends in Ecology and Evolution* 5:334–37.

———. 1995. *Interactions durables: Écologie et évolution du parasitisme.* Paris: Masson.

———. 1999. Leigh Van Valen et l'hypothèse de la Reine Rouge. In *L'évolution,* ed. H. Le Guyader, 44–52. Paris: Belin, Bibliothèque pour la Science.

———. 2000. Pressions sélectives dans les systèmes parasites-hôtes. *Journal de la Société de Biologie* 194:19–23.

———. 2001. *Parasitism: The ecology and evolution of intimate interactions.* Chicago: University of Chicago Press.

Combes, C., P. Bartoli, and A. Théron. 2002. Trematode transmission strategies. In *The Behavioural Ecology of Parasites,* ed. E. E. Lewis, J. F. Campbell, and M. V. K. Sukhdeo, 1–12. Oxford: CABI Publishing.

Combes, C., A. Fournier, H. Moné, and A. Théron. 1994. Behaviours in trematode cercariae that enhance parasite transmission: patterns and processes. *Parasitology* 109:S3–S13.

Combes, C., and A. Théron. 2000. Metazoan parasites and resource heterogenity: Constraints and benefits. *International Journal for Parasitology* 30:299–304.

Contis, G., and A. R. David. 1996. The epidemiology of *Bilharzia* in Ancient Egypt: 5000 years of Schistosomiasis. *Parasitology Today* 12:253–55.

Crofton, H. D. 1971. A model of host–parasite relationships. *Parasitology* 63:343–64.

Curie, C. R. 2001. A community of ants, fungi, and bacteria: A multilateral approach to studying symbiosis. *Annual Review of Microbiology* 55:357–80.

Davidson, D. W., and D. McKey. 1993. The evolutionary ecology of symbiotic ant–plant relationships. *Journal of Human Resources* 2:13–83.

Davies, N. B. 1988. Dumping eggs on conspecifics. *Nature* 331:19.

Davies, N. B., S. H. M. Butchart, T. A. Burke, N. Chaline, and I. R. K. Stewart. 2003. Reed warblers guard against cuckoos and cuckoldry. *Animal Behaviour* 65:285–95.

Davis, E. L., R. S. Hussey, and T. J. Baum. 2004. Getting to the roots of parasitism by nematodes. *Trends in Parasitology* 20:134–41.

Dawkins, R. 1976. *The selfish gene.* Oxford: Oxford University Press.

———. 1982. *The extended phenotype.* Oxford: Oxford University Press.

Dawkins, R., and J. R. Krebs. 1979. Arms races between and within species. *Proceedings of the Royal Society of London B* 205:489–511.

Dedeine, F., F. Vavre, F. Fleury, B. Loppin, M. E. Hochberg, and M. Boulétreau. 2001. Removing symbiotic bacteria specifically inhibits oogenesis in a parasitic wasp. *Proceedings of the National Academy of Sciences* (USA) 98:6247–52.

Denarié, J. 2000. Dialogue moléculaire des symbioses. In *Université de tous les savoirs*, vol. 1: *Qu'est-ce que la vie?* ed. Y. Michaud, 99–110. Paris: Odile Jacob.

Denison, R. F. 2000. Legume sanctions and the evolution of symbiotic cooperation by rhizobia. *American Naturalist* 156:567–76.

Despommier, D. D. 1993. *Trichinella spiralis* and the concept of niche. *Journal of Parasitology* 79:472–82.

Després, L., D. Imbert-Establet, C. Combes, and F. Bonhomme. 1992. Molecular evidence linking hominid evolution to recent radiation of schistosomes (Platyhelminthes: Trematoda). *Molecular Phylogenetics and Evolution* 1:295–304.

Després, L., D. Imbert-Establet, and M. Monnerot. 1993. Molecular characterization of mitochondrial DNA provides evidence for the recent introduction of *Schistosoma* into America. *Molecular Biochemical Parasitology* 60:221–30.

Dessein, A. J., D. Hillaire, N. E. El Wali, S. Marquet, Q. Mohamed-Ali, A. Mirghania, S. Henri, A. A. Abdelhameed, O. K. Saeed, M. M. A. Magzoub, and L. Abel. 1999. Severe hepatic fibrosis in *Schistosoma mansoni* infection is controlled by a major locus that is closely linked to the interferon-gamma receptor gene. *American Journal of Human Genetics* 65:709–21.

Dessein, A., P. Rihet, C. Demeure, P. Couissinier, O. Bacellar, E. M. Carvalho, S. Kohlstaedt, H. Dessein, A. Souza, A. Prata, V. Goudot, A. Bourgeois, and L. Abel. 1992. Facteurs génétiques et immunologiques déterminant la résistance à la bilharziose en région d'endémie. *Médecine/Sciences* 8:108–18.

Dobrovolskij, A., and G. Ataev. 2003. The nature of reproduction of trematode rediae and sporocysts. In *Taxonomy, ecology and evolution of the metazoan parasites,* ed. C. Combes and J. Jourdane, 249–72. Perpignan, France: Presses Université de Perpignan.

Dobzhansky, T. 1973. Nothing in biology makes sense except in the light of evolution. *American Biology Teacher* 35:125–29.

Doenhoff, M., R. Musallam, J. Bain, and A. McGregor. 1978. Studies on the host–parasite relationship in *Schistosoma mansoni*–infected mice: The immunological dependence of parasite egg excretion. *Immunology* 35:771–78.

Dufay, M., and M. C. Anstett. 2003. Conflicts between plants and pollinators that reproduce within inflorescences: Evolutionary variations on a theme. *Oikos* 100:3–14.

Dye, C. 2001. Fit but rare? The pros and cons of being a virulent pathogen. *Trends in Ecology and Evolution* 16:417–18.

Ebert, D., and E. A. Herre. 1996. The evolution of parasitic diseases. *Parasitology Today* 12:96–101.

Ebert, D., and K. L. Mangin. 1997. The influence of host demography on the evolution of virulence of a microsporidian gut parasite. *Evolution* 51:1828–37.

Eccles, J. C. 1989. *Evolution of the brain: Creation of the self.* London: Routledge.

Eldredge, N., and S. J. Gould. 1972. Punctuated equilibria: An alternative to phyletic gradualism. In *Models in paleobiology,* ed. T. J. M. Schopf, 82–155. San Francisco: Freeman, Cooper.

Ellis, A. G., and Midgley, J. J. 1996. A new plant–animal mutualism involving a plant with sticky leaves and a resident hemipteran insect. *Oecologia* 106:478–81.

272 Euzet, L. 1956. Recherches sur les Cestodes Tetraphyllides des Sélaciens des côtes de France. *Naturalia Monspeliensia Serie Zoologique* 3:1–265.

——. 1989. Ecologie et parasitologie. *Bulletin d'Écologie* 20:277–80.

Euzet, L., and C. Combes. 1980. Les problèmes de l'espèce chez les animaux parasites. *Bulletin de la Société Zoologique de France* 40:239–85.

——. 1998. The selection of habitats among the monogenea. *International Journal for Parasitology* 28:1645–52.

Folstad, I., and J. A. Karter. 1992. Parasites, bright males and the immunocompetence handicap. *American Naturalist* 139:603–22.

Forbes, M. L. 1966. Life cycle of *Ostrea permollis* and its relationship to the host sponge, *Stelletta grubii. Bulletin of Marine Science* 16:273–301.

Forst, S., B. Dowds, N. Boemare, and E. Stackebrandt. 1997. *Xenorhabdus* and *Photorhabdus* spp.: Bugs that kill bugs. *Annual Review of Microbiology* 51:47–72.

Frank, S. A. 2000. Polymorphism of attack and defense. *Trends in Ecology and Evolution* 15:167–71.

Galaktionov, K. V., and J. O. Bustnes. 1999. Distribution of marine bird digenean larvae in periwinkles along the southern coast of the Barents Sea. *Diseases of Aquatic Organisms* 37:221–30.

Gauld, I. D., K. J. Gaston, and D. H. Janzen. 1992. Plant allelochemicals, tritrophic interactions, and the anomalous diversity of tropical parasitoids: The "nasty" host hypothesis. *Oikos* 65:353–57.

Gehlbach, F. R., and R. S. Baldridge. 1987. Live blind snakes (*Leptotyphlops dulcis*) in eastern screech owl (*Otus asio*) nests: A novel commensalism. *Oecologia* 71:560–63.

Giberneau, M., M. Hossaert-McKey, M. C. Anstett, and F. Kjellberg. 1996. Consequences of protecting flowers in a fig: A one-way trip for pollinators? *Journal of Biogeography* 23:425–32.

Gibson, R. M., and J. W. Bradbury. 1986. Male and female mating strategies on sage grouse leks. In *Ecological aspects of social evolution*, ed. D. I. Rubenstein and R. W. Wrangham, 379–98. Princeton: Princeton University Press.

Godelle, B., and X. Reboud. 1995. Why are organelles uniparentally inherited? *Proceedings of the Royal Society of London B* 259:27–33.

Gorelick, R. 2001. Did insect pollination cause increased seed plant diversity? *Biological Journal of the Linnean Society* 74:407–27.

Gould, S. J., and R. C. Lewontin. 1979. The spandrels of San Marco and the Panglossian paradigm: A critique of the adaptationist programme. *Proceedings of the Royal Society of London B* 205:581–98.

——. 1982. L'adaptation biologique. *La Recherche* 139:1494–1502.

Gould, S. J., and E. S. Vrba. 1982. Exaptation, a missing term in the science of form. *Paleobiology* 8:4–15.

Gouyon, P.-H. 1998. Le finalisme revisité. In *L'évolution*, ed. H. Le Guyader, 40–43. Paris: Belin, Bibliothèque pour la Science.

Gouyon, P.-H., J.-P. Henry, and J. Arnould. 1997. *Les avatars du gène*. Paris: Belin.

Granovitch, A. L., S. O. Sergievsky, and L. M. Sokolova. 2000. Spatial and temporal variation of trematode infection in coexisting populations of intertidal gastropods *Littorina saxatilis* and *L. obtusata* in the White Sea. *Diseases of Aquatic Organisms* 41:53–64.

Grison-Pigé, L., J.-M. Bessière, and M. Hossaert-McKey. 2002. Specific attraction of fig-pollinating wasps: Role of volatile compounds released by tropical figs. *Journal of Chemical Ecology* 28:283–95.

Grison-Pigé, L., M. Hossaert-McKey, J. M. Greeff, and J.-M. Bessière. 2002. Fig volatile compounds—A first comparative study. *Phytochemistry* 61:61–71.

Grove, D. I. 1990. *A history of human helminthology*. London: C. A. B. International.

Gulland, F. M. D. 1992. The role of nematode parasites in Soay sheep (*Ovis aries* L.) mortality during a population crash. *Parasitology* 105:493–503.

Haag, W. R., M. L. Warren, Jr., and M. Shillingford. 1999. Host fishes and host-attracting behavior of *Lampsilis altilis* and *Villosa vibex* (Bivalvia: Unionidae). *American Midland Naturalist* 141:149–57.

Haldane, J. B. S. 1949. Disease and evolution. *Ricerca Scientifica* 19:68–76.

Hamilton, W. D., and M. Zuk. 1982. Heritable true fitness and bright birds: A role for parasites? *Science* 218:384–86.

Hanley, K. A., J. E. Biardi, C. M. Greene, T. M. Markowitz, C. E. O'Connell, and J. H. Hornberger. 1996. The behavioral ecology of host–parasite interactions: An interdisciplinary challenge. *Parasitology Today* 12:371–73.

Hasselquist, D., J. A. Marsh, P. W. Sherman, and J. C. Wingfield. 1999. Is avian humoral immunocompetence suppressed by testosterone? *Behavioral Ecology and Sociobiology* 45:167–75.

Hawthorne, D. J., and S. Via. 2001. Genetic linkage of ecological specialization and reproductive isolation in pea aphids. *Nature* 412:904–7.

Heeb, P., I. Werner, M. Kölliker, and H. Richner. 1998. Benefits of induced host responses against an ectoparasite. *Proceedings of the Royal Society of London B* 265:51–56.

Heeb, P., I. Werner, A. C. Mateman, M. Kölliker, M. W. G. Brinkhof, C. M. Lessels, and H. Richner. 1999. Ectoparasite infestation and sex-biased local recruitment of hosts. *Nature* 400:63–65.

Helluy, S. 1981. Relations hôte parasite du trématode *Microphallus papillorobustus* (Rankin, 1940). I. Pénétration des cercaires et rapports des métacercaires avec le tissu nerveux des *Gammarus* hôtes intermédiaires. *Annales de Parasitologie Humaine et Comparée* 57:263–70.

———. 1982. Relations hôte parasite du trématode *Microphallus papillorobustus* (Rankin, 1940). II. Modifications du comportement des *Gammarus* hôtes intermédiaires et localisation des métacercaires. *Annales de Parasitologie Humaine et Comparée* 58:1–17.

Herre, E. A. 1993. Population structure and the evolution of virulence in nematode parasites of fig wasps. *Science* 259:1442–45.

———. 1995. Factors affecting the evolution of virulence: Nematode parasites of fig wasps as a case study. *Parasitology* 111: S179–S191.

274

Hoberg, E. P., N. L. Alkire, A. de Queiroz, and A. Jones. 2001. Out of Africa: Origins of the *Taenia* tapeworms in humans. *Proceedings of the Royal Society of London B* 268:781–87.

Hoberg, E. P., A. Jones, R. L. Rausch, K. S. Eom, and S. L. Gardner. 2000. A phylogenetic hypothesis for species of the genus *Taenia* (Eucestoda: Taeniidae). *Journal of Parasitology* 86:89–98.

Hochberg, M. E. 1997. Hide or fight? The competitive evolution of concealment and encapsulation in parasitoid–host associations. *Oikos* 80:342–52.

Holmes, J. C. 1989. A redescription of *Simhatrema simhai* Chattopadhyaya, 1970 (Trematoda: Exotidendriidae), with comments on its pathogenesis in sea snakes (Serpentes: Hydrophiidae). *Proceedings of the Helminthological Society of Washington* 56:156–61.

Hooper, J. 2002. *Of moths and men: An evolutionary tale.* New York: Norton.

Hourdry, J., P. Cassier, J.-L. D'Hondt, and M. Porchet. 1995. *Métamorphoses animales: Transitions écologiques.* Paris: Hermann.

Jacob, F. 1977. Evolution and tinkering. *Science* 196:1161–66.

Jaisson, P. 1993. *La fourmi et le sociobiologiste.* Paris: Odile Jacob.

James, E. R., and D. R. Green. 2004. Manipulation of apoptosis in the host–parasite interaction. *Trends in Parasitology* 20:280–86.

Johnson, K. P., R. J. Adams, R. D. Page, and D. H. Clayton. 2003. When do parasites fail to speciate in response to host speciation? *Systematic Biology* 52:37–47.

Johnson, L. L., and M. S. Boyce. 1991. Female choice of males with low parasite loads in sage grouse. In *Bird–parasite interactions: Ecology, evolution, and behaviour,* ed. J. E. Loye and M. Zuk, 377–88. Oxford: Oxford University Press.

Jourdane, J. 1974. Découverte de l'hôte vecteur de *Nephrotrema truncatum* (Leuckart, 1842) (Trematoda) et mise en évidence d'une phase hépatique au cours de la migration du parasite chez l'hôte définitif. *Comptes Rendus de l'Académie des Sciences D* 278:1533–36.

Jourdane, J., and A. Théron. 1987. Larval development: Eggs to cercariae. In *The biology of schistosomes,* ed. D. Rollinson and A. J. G. Simpson, 83–113. London: Academic Press.

Jousselin, E., J.-Y. Rasplus, and F. Kjellberg. 2001. Shift to mutualism in parasitic lineages of the fig/fig wasp interaction. *Oikos* 94:287–94.

Jousson, O., P. Bartoli, and J. Pawlowski. 2000. Cryptic speciation among intestinal parasites (Trematoda: Digenea) infecting sympatric host fishes. *Journal of Evolutionary Biology* 13:778–85.

Jovani, R. 2003. Understanding parasite strategies. *Trends in Parasitology* 19:15–16.

Kahn, A. 2000. *Et l'homme dans tout ça.* Paris: Editions Nil.

Kavaliers, M., and D. Colwell. 1994. Parasite infection attenuates nonopioid mediated predator-induced analgesia in mice. *Physiology and Behaviour* 55:505–10.

———. 1995. Decreased predator avoidance in parasitized mice: Neuromodulatory correlates. *Parasitology* 111:257–63.

Kechemir, N. 1978. Démonstration expérimentale d'un cycle à quatre hôtes obligatoires chez les trématodes hémiurides. *Annales de Parasitologie Humaine et Comparée* 53:75–92.

Kerr, P. J., and S. M. Best. 1998. *Myxoma* virus in rabbits. *Revue Scientifique et Technique de l'Office International des Epizooties* 17:256–68.

Kiers, E. T., R. A. Rousseau, S. A. West, and R. F. Denison. 2003. Host sanctions and the legume–rhizobium mutualism. *Nature* 425:78–81.

Kjellberg, F., P.-H. Gouyon, M. Ibrahim, M. Raymond, and G. Valdeyron. 1987. The stability of the symbiosis between dioecious figs and their pollinators: A study of *Ficus carica* L. and *Blastophaga psenes* L. *Evolution* 41:693–704.

Klein, R. G. 1989. *The human career: Human biological and cultural origins.* Chicago: University of Chicago Press.

Ladevèze, V., S. Aulard, N. Chaminade, G. Périquet, and F. Lemeunier. 1998. Hobo transposons causing chromosomal breakpoints. *Proceedings of the Royal Society of London B* 265:1157–59.

Lafferty, K. D., and A. K. Morris. 1996. Altered behavior of parasitized killifish greatly increases susceptibility to predation by bird final hosts. *Ecology* 77:1390–97.

Laland, K. N., F. J. Odling-Smee, and M. W. Feldman. 1996. On the evolutionary consequences of niche construction. *Journal of Evolutionary Biology* 9:293–316.

———. 2000. Niche construction, biological evolution, and cultural change. *Behavioral and Brain Sciences* 23:131–75.

Langmore, N. E., S. Hunt, and R. M. Kilner. 2003. Escalation of a coevolutionary arms race through host rejection of brood parasitic young. *Nature* 422:157–60.

Lashley, F. R., and J. D. Durham. 2002. *Emerging infectious diseases: Trends and issues.* New York: Springer.

Leigh, E. G. 1971. *Adaptation and diversity.* San Francisco: Freeman, Cooper.

Leigh, E. G., and T. Rowell. 1995. The evolution of mutualism and other forms of harmony at various levels of biological organization. *Écologie* 26:131–58.

Leroy, E. M., P. Rouquet, P. Formenty, S. Souquière, A. Kilbourne, J.-M. Froment, M. Bermejo, S. Smit, W. Karesh, R. Swanepoel, S. R. Zaki, and P. E. Rollin. 2004. Multiple Ebola virus transmission events and rapid decline of central African wildlife. *Science* 303:387–90.

Levin, B. R. 1996. The evolution and maintenance of virulence in microparasites. *Emerging Infectious Diseases* 2:93–102.

Li, H. G., T. Fujiyoshi, H. Lou, S. Yahiki, S. Sonoda, L. Cartier, L. Nunez, I. Munoz, S. Horai, and K. Tajima. 1999. The presence of ancient human T-cell lymphotrophic virus type I provirus DNA in an Andean mummy. *Nature Medicine* 5:1428.

Lindholm, A. K. 1999. Brood parasitism by the cuckoo on patchy reed warbler populations in Britain. *Journal of Animal Ecology* 68:293–309.

Lockyer, A. E., P. D. Olson, P. Ostergaard, D. Rollinson, D. A. Johnston, S. W. Attwood, V. R. Southgate, P. Horak, S. D. Snyder, T. H. Le, T. Agatsuma, D. P. McManus, A. C. Carmichael, S. Naem, and D. T. J. Littlewood. 2003. The phylogeny of the Schistosomatidae based on three genes with emphasis on the interrelationships of *Schistosoma* Weinland, 1858. *Parasitology* 126:203–24.

Lyon, B. E. 2003. Egg recognition and counting reduce costs of avian conspecific brood parasitism. *Nature* 422:495–99.

276   Machado, C. A., E. Jousselin, F. Kjellberg, S. G. Compton, and E. A. Herre. 2001. Phylogenetic relationships, historical biogeography and character evolution of fig-pollinating wasps. *Proceedings of the Royal Society of London B* 268:685–94.

Maillard, C. 1976. Distomatoses de poissons en milieu lagunaire. Ph.D. thesis, Université de Montpellier II, France.

Majerus, M. E. N. 2003. Industrial melanism. In *Encyclopedia of insects,* ed. V. H. Resh and R. T. Cardé, 560–64. San Diego: Academic Press.

Marchetti, K., H. Nakamura, and H. Lisle Gibbs. 1998. Host-race formation in the common cuckoo. *Science* 282:471–72.

Margulis, L. 1992. *Symbiosis in cell evolution.* Oxford: Freeman.

Marr, J. S., and C. H. Calisher. 2003. Alexander the Great and West Nile virus encephalitis. *Emerging Infectious Diseases* 9:1599–1603.

Martin, P. M. V., and C. Combes. 1996. Emerging infectious diseases and the depopulation of French Polynesia in the nineteenth century. *Emerging Infectious Diseases* 2:359–61.

Massoud, Z. 1992. *Terre vivante.* Paris: Odile Jacob.

Maynard Smith, J. 1989. Generating novelty by symbiosis. *Nature* 341:384–85.

Maynard Smith, J., and E. Szathmary. 1997. *The major transitions in evolution.* Oxford: Oxford University Press.

Mayr, E. 1999. Understanding evolution. *Trends in Ecology and Evolution* 14:372–73.

McKeown, T. 1988. *The origins of human diseases.* Oxford: Blackwell.

McKerrow, J. H. 1997. Cytokine induction and exploitation in schistosome infections. *Parasitology* 115: S107–S112.

Meinesz, A. 1999. *Killer algae.* Chicago: University of Chicago Press.

Mesoudi, A., A. Whiten, and K. N. Laland. 2004. Perspective: Is human cultural evolution Darwinian? Evidence reviewed from the perspective of *The Origin of Species. Evolution* 58:1–11.

Meyhöfer, R., and J. Casas. 1999. Vibratory stimuli in host location by parasitic wasps. *Journal of Insect Physiology* 45:967–71.

Meyhöfer, R., J. Casas, and S. Dorn. 1997. Vibration mediated interactions in a host–parasitoid system. *Proceedings of the Royal Society of London B* 264:261–66.

Mitchell, C. E., and A. G. Power. 2003. Release of invasive plants from fungal and viral pathogens. *Nature* 421:625–27.

Modiano, D., G. Luoni, B. S. Sirima, J. Simpore, F. Verra, A. Konate, E. Rastrelli, A. Olivieri, C. Calissano, G. M. Paganotti, L. D'Urbano, I. Sanou, A. Sanadogo, G. Modiano, and M. Coluzzi. 2001. Haemoglobin C protects against clinical *Plasmodium falciparum* malaria. *Nature* 414:305–8.

Møller, A. P. 1988. Female choice selects for male sexual tail ornaments in the monogamous swallow. *Nature* 332:640–41.

Møller, A. P., P. Christe, and E. Lux. 1999. Parasitism, host immune function, and sexual selection. *Quarterly Review of Biology* 74:3–20.

Møller, A. P., R. Dufva, and J. Erritzoe. 1998. Host immune function and sexual selection in birds. *Journal of Evolutionary Biology* 11:703–19.

Moore, J. 1993. Worthy animals. *Science* 262:124.

———. 1995. The behavior of parasitized animals—When an ant is not an ant. *BioScience* 45:89–96.

Moore, J. C., A. M. Dunn, S. G. Compton, and M. J. Hatcher. 2003. Foundress re-emergence and fig permeability in fig tree–wasp mutualisms. *Journal of Evolutionary Biology* 16:1186–95.

Moore, S. L., and K. Wilson. 2002. Parasites as a viability cost of sexual selection in natural populations of mammals. *Science* 297:2015–18.

Mooring, M. S., and P. J. Mundy. 1996. Factors influencing host selection by yellow-billed oxpeckers at Matobo National Park, Zimbabwe. *African Journal of Ecology* 34:177–88.

Morand, S. 1996. Life history traits in parasitic nematodes: A comparative approach for the search of invariants. *Functional Ecology* 10:210–18.

Morand, S., P. Legendre, S. L. Gardner, and J.-P. Hugot. 1996. Body size evolution of oxyurid (Nematoda) parasites: The role of hosts. *Oecologia* 107:274–82.

Morand, S., F. Robert, and V. Connors. 1995. Complexity in parasite life-cycles: Population biology of cestodes in fish. *Journal of Animal Ecology* 64:256–64.

Munger, J. C., and J. C. Holmes. 1988. Benefits of parasitic infection: A test using a ground squirrel–trypanosome system. *Canadian Journal of Zoology* 66:222–27.

Musser, J. 1996. Molecular population genetic analysis of emerged bacterial pathogens: Selected insights. *Emerging Infectious Diseases* 2:1–17.

Nenon, J.-P., N. Kacem, and J. Le Lannic. 1997. Structure, sensory equipment, and secretions of the ovipositor in a giant species of *hymenoptera, Megarhyssa atrata* F. (Ichneumonidae, Pimplinae). *Canadian Entomologist* 129:789–99.

Newhouse, J. R. 1990. Chestnut blight. *Scientific American* 263:106–11.

Nilsson, L. A. 1998. Deep flowers for long tongues. *Trends in Ecology and Evolution* 13:259–60.

Odling-Smee, F. J., K. N. Laland, and M. W. Feldman. 2003. *Niche construction: The neglected process in evolution.* Princeton: Princeton University Press.

Olivieri, I., and S. A. Frank. 1994. The evolution of nodulation in *Rhizobium*: Altruism in the rhizosphere. *Journal of Heredity* 85:46–47.

Olson, P. D., D. T. Littlewood, D. Griffiths, C. R. Kennedy, and C. Arme. 2002. Evidence for the co-existence of separate strains or species of *Ligula* in Lough Neagh, Northern Ireland. *Journal of Helminthology* 76:171–74.

Oppliger, A., P. Christe, and H. Richner. 1996. Clutch size and malaria resistance. *Nature* 381:565.

Oppliger, A., H. Richner, and P. Christe. 1994. Effect of an ectoparasite on lay date, nest-site choice, desertion, and hatching success in the great tit (*Parus major*). *Behavioral Ecology* 5:130–34.

Orgel, L. E., and F. H. C. Crick. 1980. Selfish DNA: The ultimate parasite. *Nature* 284:604–7.

Pagano, A. 1999. Les complexes hybridogénétiques de grenouilles vertes: Déterminants de la distribution dans la vallée alluviale du Rhône. Ph.D. thesis, Université Lyon I, France.

Page, R. D. M., P. L. M. Lee, S. A. Becher, R. Griffiths, and D. H. Clayton. 1998. A different tempo of mitochondrial DNA evolution in birds and their parasitic lice. *Molecular Phylogenetics and Evolution* 9:276–93.

Pagès, J.-R., and A. Théron. 1990. Analysis and comparison of cercarial emergence rhythms of *Schistosoma haematobium, S. intercalatum* and *S. bovis* and their hybrid progeny. *International Journal for Parasitology* 20:193–97.

Pampoulie, C., A. Lambert, E. Rosecchi, A. J. Crivelli, J. L. Bouchereau, and S. Morand. 2000. Host death: A necessary condition for the transmission of *Aphalloides coelomicola* (Dollfus, Chabaud, and Golvan, 1957) (Digenea, Cryptogonimidae)? *Journal of Parasitology* 86:416–17.

Pariselle, A. 1997. Diversité, spéciation et évolution des monogènes branchiaux de Cichlidae en Afrique de l'Ouest. Ph.D. thesis, Université de Perpignan, France.

Passera, L. 1975. Les fourmis hôtes provisoires ou intermédiaires des helminthes. *Année Biologique* 14:227–57.

Pennisi, E. 2003. On ant farm, a threesome evolves. *Science* 299:325.

Péru, L. 1982. Fourmis du genre *Leptothorax* et cestodes cyclophyllides: Modifications de l'hôte intermédiaire sous l'influence du cysticercoïde. Ph.D. thesis, Université Paris 6.

Picot, H., and J. Benoist. 1975. Interaction of social and ecological factors in the epidemiology of helminth parasites. In *Biosocial interrelations in population adaptation*, ed. E. Watts, 233–47. The Hague: Mouton.

Pigliucci, M., and J. Kaplan. 2000. The fall and rise of Dr. Pangloss: Adaptationism and the spandrels paper 20 years later. *Trends in Ecology and Evolution* 15:66–70.

Plateaux, L. 1972. Sur les modifications produites chez une fourmi par la présence d'un parasite cestode. *Annales des Sciences Naturelles Zoologie et Biologie Animale* 14:203–20.

Poulin, R. 1992a. Altered behaviour in parasitized bumblebees: Parasite manipulation or adaptive suicide? *Animal Behaviour* 44:174–76.

———. 1992b. Determinants of host-specificity in parasites of freshwater fishes. *International Journal for Parasitology* 22:753–58.

———. 1995. Adaptive changes in the behaviour of parasitized animals: A critical review. *International Journal for Parasitology* 25:1371–83.

———. 1998. *Evolutionary ecology of parasites*. London: Chapman and Hall.

Poulin, R., and C. Combes. 1999. The concept of virulence: Interpretations and implications. *Parasitology Today* 15:474–75.

Poulin, R., and W. L. Vickery. 1993. Parasite distribution and virulence: Implications for parasite-mediated sexual selection. *Behavioral Ecology and Sociobiology* 33:429–36.

Prévot, G., P. Bartoli, and S. Deblock. 1976. Cycle biologique de *Maritrema misenensis* (A. Palombi, 1940) n. comb. (Trematoda: Microphallidae Travassos, 1920) du Midi de la France. *Annales de Parasitologie Humaine et Comparée* 51:433–46.

Price, P. W. 1980. *Evolutionary biology of parasites*. Princeton: Princeton University Press.

———. 1989. The web of life: Development over 3.8 billion years of trophic relationships. In *Symbiosis as a source of evolutionary innovation,* ed. L. Margulis and R. Fester, 262–72. Cambridge: MIT Press.

Raikova, E. V. 1994. Life cycle, cytology, and morphology of *Polypodium hydriforme*, a coelenterate parasite of the eggs of Acipenseriform fishes. *Journal of Parasitology* 80:1–22.

Rasplus, J.-Y., C. Kerdelhué, I. Le Clainche, and G. Mondor. 1998. Molecular phylogeny of fig wasps: Agaonidae are not monophyletic. *Comptes Rendus de l'Académie des Sciences*, series 3, *Sciences de la Vie* 321:517–27.

Read, A. F. 1988. Sexual selection and the role of parasites. *Trends in Ecology and Evolution* 3:97–101.

Read, A. F., and A. Skorping. 1995. The evolution of tissue migration by parasitic nematode larvae. *Parasitology* 111:359–71.

Read, A. P., and L. H. Taylor. 2001. The ecology of genetically diverse infections. *Science* 292:1099–1102.

Richner, H., P. Christe, and A. Oppliger. 1995. Paternal investment affects prevalence of malaria. *Proceedings of the National Academy of Sciences* (USA) 92:1192–94.

Richner, H., and P. Heeb. 1995. Are clutch and brood size patterns in birds shaped by ectoparasites? *Oikos* 73:435–41.

Richner, H., A. Oppliger, and P. Christe. 1993. Effect of an ectoparasite on the reproduction in great tits. *Journal of Animal Ecology* 62:703–10.

Richner, H., and F. Tripet. 1999. Ectoparasitism and the trade-off between current and future reproduction. *Oikos* 86:535–38.

Ridley, M. 1993. *Evolution*. Oxford: Blackwell Scientific Publications.

———. 1994. *The Red Queen: Sex and the evolution of human nature*. London: Penguin Books.

Robert, M., and G. Sorci. 1999. Rapid increase of host defence against brood parasites in a recently parasitized area: The case of village weavers in Hispaniola. *Proceedings of the Royal Society of London B* 266:941–46.

Rodhain, F. 2002. Mécanismes de transmission des maladies vectorielles. *Environnement, Risques et Santé* 1 (special issue 1): 532–34.

Rohde, K. 1994. Niche restriction in parasites: Proximal and ultimate causes. *Parasitology* 109: S69–S84.

Rosenqvist, G., and K. Johansson. 1995. Male avoidance of parasitized females explained by direct benefits in a pipefish. *Animal Behaviour* 49:1039–45.

Rothstein, S. I. 1990. Brood parasitism and clutch size determination in birds. *Trends in Ecology and Evolution* 5:101–2.

Rothstein, S. I., and S. K. Robinson. 1994. Conservation and coevolutionary implications of brood parasitism by cowbirds. *Trends in Ecology and Evolution* 9:162–64.

Roy, B. A. 1994. The use and abuse of pollinators by fungi. *Trends in Ecology and Evolution* 9:335–38.

Sasal, P., E. Jobet, E. Faliex, and S. Morand. 2000. Sexual competition in an acanthocephalan parasite of fish. *Parasitology* 120:65–69.

Schluter, D. 2001. Ecology and the origin of species. *Trends in Ecology and Evolution* 16:372–80.

Selosse, M.-A., B. Albert, and B. Godelle. 2001. Reducing the genome size of organelles favours gene transfer to the nucleus. *Trends in Ecology and Evolution* 16:136–41.

280   Selosse, M.-A., and S. Loiseau-de Goër. 1997. La saga de l'endosymbiose. *La Recherche* 296:36–41.

Sigmund, K., and C. Hauert. 2002. Primer: Altruism. *Current Biology* 12 (8): R270–R272.

Skorping, A., A. F. Read, and A. E. Keymer. 1991. Life history covariation in intestinal nematodes of mammals. *Oikos* 60:365–72.

Smith, D. C. 1992. The symbiotic condition. *Symbiosis* 14:3–15.

Smith, D. C., and A. E. Douglas. 1987. *The biology of symbiosis*. Baltimore: Edward Arnold.

Smith, S. E., and D. J. Read. 1997. *Mycorrhizal symbiosis*. San Diego: Academic Press.

Smyth, J. D. 1994. *Introduction to animal parasitology*. Cambridge: Cambridge University Press.

Snyder, S. D., and E. S. Loker. 2000. Evolutionary relationships among the Schistosomatidae (Platyhelminthes: Digenea) and an Asian origin for *Schistosoma*. *Journal of Parasitology* 86:283–88.

Sorvillo, F., L. R. Ash, O. G. Berlin, and S. A. Morse. 2002. *Baylisascaris procyonis*: An emerging helminthic zoonosis. *Emerging Infectious Diseases* 8:355–59.

Southgate, V. R., D. S. Brown, A. Warlow, R. J. Knowles, and A. Jones. 1989. The influence of *Calicophoron microbothrium* on the susceptibility of *Bulinus* tropicus to *Schistosoma bovis*. *Parasitology Research* 75:381–91.

Starr, D., and T. W. Cline. 2002. A host–parasite interaction rescues *Drosophila* oogenesis defects. *Nature* 418:76–79.

Stenseth, N. C., and J. Maynard Smith. 1984. Coevolution in ecosystems: Red Queen evolution or stasis? *Evolution* 38:870–80.

Storfer, A., and A. Sih. 1998. Gene flow and ineffective antipredator behavior in a stream-breeding salamander. *Evolution* 52:558–65.

Sukhdeo, M. K. V. 2000. Inside the vertebrate host: Ecological strategies by parasites living in the third environment. In *Evolutionary biology of host–parasite relationships: Theory meets reality*, ed. R. Poulin, S. Morand, and A. Skorping, 43–62. Amsterdam: Elsevier.

Sures, B. 2004. Environmental parasitology: Relevancy of parasites in monitoring environmental pollution. *Trends in Parasitology* 20:170–77.

Sures, B., R. Siddall, and H. Taraschewski. 1999. Parasites as accumulation indicators of heavy metal pollution. *Parasitology Today* 15:16–21.

Taddei, F., I. Matic, and M. Radman. 2000. SOS génome: Réparation et évolution. *Pour la Science* 269:66–73.

Taylor, D. R., A. M. Jarosz, R. E. Lenski, and D. W. Fulbright. 1998. The acquisition of hypovirulence in host–pathogen systems with three trophic levels. *American Naturalist* 151:343–55.

Temime, L., P. Y. Boëlle, P. Courvalin, and D. Guillemot. 2003. Bacterial resistance to penicillin G by decreased affinity of penicillin-binding proteins: A mathematical model. *Emerging Infectious Diseases* 9:411–17.

Théron, A. 1984. Early and late shedding patterns of *Schistosoma mansoni* cercariae: Ecological significance in transmission to human and murine hosts. *Journal of Parasitology* 70:652–55.

Théron, A., and C. Combes. 1988. Genetic analysis of cercarial emergence rhythms of *Schistosoma mansoni*. *Behavior Genetics* 18:201–9.

———. 1995. Asynchrony of infection timing, habitat preference and sympatric speciation of schistosome parasites. *Evolution* 49:372–75.

Théron, A., C. Sire, A. Rognon, F. Prugnolle, and P. Durand. 2004. Molecular ecology of *Schistosoma mansoni* transmission inferred from the genetic composition of larval and adult infrapopulations within intermediate and definitive hosts. *Parasitology*, in press.

Toft, C. A. 1991. Current theory of host–parasite interactions. In *Bird–parasite interactions: Ecology, evolution, and behaviour*, ed. J. E. Loye and M. Zuk, 3–15. Oxford: Oxford University Press.

Tompkins, D. M., and D. H. Clayton. 1999. Host resources govern the specificity of swiftlet lice: Size matters. *Journal of Animal Ecology* 68:489–500.

Torchin, M. E., K. D. Lafferty, A. P. Dobson, V. J. McKenzie, and A. M. Kuris. 2003. Introduced species and their missing parasites. *Nature* 421:628–30.

Tripet, F., and H. Richner. 1997a. The coevolutionary potential of a "generalist" parasite, the hen flea *Ceratophyllus gallinae*. *Parasitology* 115:419–27.

———. 1997b. Host responses to ectoparasites: Food compensation by parent blue tits. *Oikos* 78:557–61.

Valkiunas, G., and T. Iezhova. 2000. A key to some puzzles in ecological parasitology. *Ekologiya* (Vilnius) 2:27–30.

Van Valen, L. 1973. A new evolutionary law. *Evolution Theory* 1:1–30.

Vavre, F., F. Fleury, D. Lepetit, P. Fouillet, and M. P. Boulétreau. 1999. Phylogenetic evidence for horizontal transmission of *Wolbachia* in host–parasitoid associations. *Molecular Biology and Evolution* 16:1711–23.

Volkman, S. K., A. E. Barry, E. J. Lyons, K. M. Nielsen, S. M. Thomas, M. Choi, S. S. Thakore, K. P. Day, F. D. Wirth, and D. Hartl. 2001. Recent origin of *Plasmodium falciparum* from a single progenitor. *Science* 293:482–84.

Wasserthal, L. T. 1997. The pollinators of the Malagasy star orchids *Angraecum sesquipedale*, *A. sororium*, and *A. compactum* and the evolution of extremely long spurs by pollinator shift. *Botanica Acta* 110:343–49.

Westneat, D. F., D. Hasselquist, and J. C. Wingfield. 2003. Tests of association between the humoral immune response of red-winged blackbirds (*Agelaius phoeniceus*) and male plumage, testosterone, or reproductive success. *Behavioral Ecology and Sociobiology* 53:315–23.

Westphall, E., R. Bronner, and P. Michler. 1987. *Découvrir et reconnaître les galles*. Neuchâtel, Switzerland: Delachaux et Niestlé.

Whittington, I. D. 1996. Benedeniine capsalid monogeneans from Australian fishes: Pathogenic species, site-specificity and camouflage. *Journal of Helminthology* 70:177–84.

Willmer, P. G., and G. N. Stone. 1997. How aggressive ant-guards assist seed-set in *Acacia* flowers. *Nature* 388:165–67.

Winfree, R. 1999. Cuckoos, cowbirds and the persistence of brood parasitism. *Trends in Ecology and Evolution* 14:338–43.

282   Woese, C. 2002. On the evolution of cells. *Proceedings of the National Academy of Sciences* (USA) 99:8742–47.

Wolowczuk, I., S. Nutten, O. Roye, M. Delacre, M. Capron, R. M. Murray, F. Trottein, and C. Auriault. 1999. Infection of mice lacking interleukin-7 (IL-7) reveals an unexpected role for IL-7 in the development of the parasite *Schistosoma mansoni*. *Infection and Immunity* 67:4183–90.

Wolowczuk, I., O. Roye, S. Nutten, M. Delacre, F. Trottein, and C. Auriault. 1999. Role of interleukin-7 in the relation between *Schistosoma mansoni* and its definitive vertebrate host. *Microbes and Infection* 7:545–51.

Woltz, P., R. A. Stockey, M. Gondran, and J. F. Cherrier. 1994. Interspecific parasitism in the Gymnosperms: Unpublished data on two endemic New Caledonian Podocarpaceae using scanning electron microscopy. *Acta Botanica Gallica* 141: 731–46.

Xia, M., J. Jourdane, and C. Combes. 1998. Local adaptation of *Schistosoma japonicum* in its snail host demonstrated by transplantation of sporocysts. In *Proceedings of the Ninth International Congress of Parasitology*, ed. I. Tada, S. Kojima, and M. Tsuji, 573–76. Bologne, Italy: Monduzzi Editore.

Young, J. P. W., and A. W. B. Johnston. 1989. The evolution of specificity in the legume–*Rhizobium* symbiosis. *Trends in Ecology and Evolution* 4:341–49.

Zahavi, A., and A. Zahavi. 1997. *The handicap principle*. Oxford: Oxford University Press.

INDEX